Water Technology in the Middle Ages

THE JOHNS HOPKINS STUDIES
IN THE HISTORY OF TECHNOLOGY

Merritt Roe Smith, Editor

Cities, Monasteries, and Waterworks after the Roman Empire

Water Technology in the Middle Ages

Roberta J. Magnusson

The Johns Hopkins University Press
Baltimore and London

This book has been brought to publication with the generous assistance of the Department of History, University of Oklahoma.

Printed in the United States of America
on acid-free paper

9 8 7 6 5 4 3 2 1

The Johns Hopkins University Press
2715 North Charles Street
Baltimore, Maryland 21218-4363
www.press.jhu.edu

A catalog record for this book is available from the British Library.

Library of Congress Cataloging-in-Publication Data

Magnusson, Roberta J., 1952–
Water technology in the Middle Ages : cities, monasteries, and waterworks after the Roman Empire / Roberta J. Magnusson.
p. cm. — (Johns Hopkins studies in the history of technology)
Includes bibliographical references and index.
ISBN 0-8018-6626-X
1. Water-supply engineering—Europe—History. 2. Municipal water supply—Europe—History. 3. Monasteries—Europe—Water-supply—History. 4. Waterworks—Europe—History. 5. Middle Ages—History. I. Title. II. Series.

TC455.M34 2001
627′.094′0902—DC21
00-011509

Contents

Preface

As the sun rose one Monday morning in the spring of 1339, twelve-year-old Ralph de Mymmes, a groom employed by carter John Absolon, was driving his master's cart through the streets of London. The cart, drawn by a pair of horses, carried a cask full of water. As the boy drove along Chepe, seven-year-old John le Stolere, "a pauper and mendicant," was squatting in the street to relieve himself. Perhaps Ralph was still sleepy, or perhaps he simply did not see the small boy in the road. Whatever the reason, John was accidentally crushed under one of the wheels of the cart, and he died instantly. The young driver took fright and fled, abandoning his master's water-cart and horses—he had still not been found at the time the coroner recorded the death.[1]

This unfortunate accident encapsulates the collision of two fundamental issues facing the communities of medieval Europe: water supplies and waste disposal. Did a beggar boy have no better alternative than to relieve himself amid the traffic in the street? Had Ralph filled his water cask at one of the docks at the riverfront, where the Thames was polluted with the sewage generated by John and thousands of other Londoners? Or had Ralph stopped to fill his cask that morning with clean piped water from London's nearby Great Conduit?

Hydraulic engineering stands at the interface between human needs and the natural world. A society's ability to harness its hydraulic resources reveals something of its ability to control the natural environment as well as its ability to organize its members. Admittedly, advanced hydraulic engineering is not usually the first thing to come to mind when one thinks about the Middle Ages. My own interest in the subject owes much to my bottom-up introduction to medieval civilization, based on years spent as an itinerant field archaeologist. While I was a graduate student at Berkeley, a history seminar turned to the topic of the Cistercians. Each of us was asked to say what our first mental image was when we thought about the famous monastic order. Having grubbed all too intimately in medieval cesspits, my reply came from the heart: "Great drains!" This response was greeted with baffled amusement by my fellow students, whose refined minds turned more naturally to affective spirituality and other such elevated themes. Yet it was Saint Bernard himself who praised the Virgin Mary by comparing her to an aqueduct and who was reluctant to move to a larger site when droves of enthusiastic new recruits were crowding into the first monastic buildings at Clairvaux, on the grounds that the fledgling community would be maligned as frivolous and unstable were they to abandon their expensive new water system.[2]

In this study I take a close look at the interrelationship between people and one technology cluster: complex, gravity-flow water systems. Such systems, which were composed of collection basins, long-distance conduits, and distribution points, were built for medieval palaces, castles, manors, hospitals, gardens, and at least one enterprising village. I focus primarily on the two most common types of large-scale water systems: monastic and urban conduits. Why, how, and by whom were they built? How well did they work? Who used them, and for what purposes? How were they paid for? How were they maintained? How common were they? At a more general level, what impact did hydraulic technology have on medieval society, and what impact did medieval society have on hydraulic technology?

Like many other "medieval" technologies, gravity-flow hydraulic engineering was not an entirely new invention. Medieval Europe had inherited a highly developed range of Roman hydraulic components. The basic technological trajectory, based on low-pressure systems of channels and pipes, was already established. Individual medieval components, such as pipes and taps, were often nearly identical to their Roman counterparts. Nonetheless, it is not enough to dismiss medieval hydraulic engineering as merely derivative. Technologies are not immutable, nor are technological trajectories permanently fixed. The medieval pipes and channels unearthed by the archaeologist are not merely physical objects; they are also cultural artifacts. The apparent resemblance between a medieval lead pipe and a Roman lead pipe is the end result of a historical process, which requires a historical explanation. Was it the product of a continuous technological tradition, and if so, how was that tradition transmitted over the centuries? Or was Roman-style engineering rediscovered and revived after a technological hiatus?

That pipe, moreover, was culturally embedded in medieval society. It was fashioned by medieval craftsmen, using materials obtained and transported in the Middle Ages, as part of an overall hydraulic system built for a particular purpose and for a specific sponsor. Its pipe trench cut through land held according to medieval patterns of tenure, it was paid for according to medieval standards of wages and prices, in money raised by medieval forms of financial exactions. It delivered water to a structure used by medieval men and women, who employed the water for a range of culturally specific activities. In short, however Roman its physical appearance may be, its full significance can be understood only within the context of its own society.

If our pipe was the outcome of a set of specifically medieval contingencies and trade-offs, can it be of any use to broader historical questions or more generalized theories of technological change? I believe that it can, inasmuch as medieval water systems have the potential to provide a longitudinal case study for the evolution of technological systems in a premodern society. Unlike some other medieval technologies, such as stirrups or horseshoes, water systems can be documented and dated with a fair degree of confidence. The conduits of the High Middle Ages were the products of a society that was increasingly literate and kept extensive records. References to water appear in a wide variety of sources, such as charters, administrative and financial records, court records, and law codes. Furthermore, many of the physical components of water systems were subterranean structures, which have a good chance of surviving archaeologically, or at least of leaving recognizable traces in datable contexts. In spite of the lack of a systematic medieval treatise on hydraulic engineering, there is abundant (if often widely scattered) evidence for the technology's "human components" and social context. Craftsmen, sponsors, wardens, repairmen, users—even malefactors and accident victims—appear as named individuals interacting with water systems.

Theoretical models of technological change can help the historian make sense of the medieval evidence, though the application of models derived from the modern world can be problematic. The incomplete and patchy nature of the medieval sources is the most immediate stumbling block. It is not possible to compile a definitive list of medieval water systems. Even a complete corpus of textual references, should one ever become available, would be insufficient: archaeologists continue to provide evidence for water systems that have completely eluded documentary historians. We are forced to rely on a nonrandom sample, without knowing exactly how skewed it is or how great a proportion of the whole it represents. Even for known hydraulic systems, the evidence is very uneven. Some stages in the evolution of a system, such as land acquisition, left an abundant paper trail, whereas others left little or no documentation or physical evidence. That the surviving sources preserve only a partial picture becomes particularly evident in the rare cases in which a water system is relatively well documented. The author of the sixteenth-century *Rites of Durham* presents a sunny image of monks washing at the laver fountain and drying their hands on clean towels before entering the refectory. The Durham account rolls, on the other hand, paint a far grimmer picture of women water-carriers hauling water up the steep hill from the Wear River because the abbey's pipes were once again fractured or frozen. Archaeological excavations have confirmed the architectural details presented by the author of the *Rites* but also suggest that the pipe trenches were too shallow to protect the pipes from frost damage.[3]

Like modern technological systems, medieval water systems were not simply composed of physical artifacts. The networks of interconnected hydraulic components that constituted water systems in the technical sense were inextricably linked to specific physical and social environments. In the broader sense of the term, their systems included natural resources, organizational structures, laws, technical knowledge (and mechanisms for its transmission), and users. All of these elements had to be successfully combined for a hydraulic system to operate effectively.

The evolution of medieval technology also suggests some instructive contrasts with modern patterns of technological change. The water systems that form the basis of this study were not initiated by central government policies, capitalist entrepreneurs, or professional, externally sponsored "change agents." They were, rather, the inexpert, experimental, trial-and-error responses of local communities to locally recognized needs. Furthermore, the rate of technology transfer and technical change was slow: a chronological scale of centuries, not years or decades, is required.[4]

As a case study, the analysis of hydraulic technology can contribute to our understanding of medieval society. Was medieval Europe unusually innovative and inventive, a society that eagerly adopted and developed technical innovations? Or was it more resistant to technical change? How did it transmit and adapt technical traditions? What role did religion or religious institutions play in technological development? Why were the advantageous technical methods and techniques that were known in the Middle Ages not more widely employed?

A few preliminary remarks about the scope and limitations of the present work may be helpful. Rather than to present the reader with a series of fragmentary local studies, I have focused on different stages in the evolution of medieval water systems by bringing together comparative information from multiple sites. The geographic parameters cover Western Christendom from Italy to Scandinavia, except for Christian Spain (which was heavily influenced by Islamic technology and practices and seems to have stood somewhat apart from the rest of Europe in its hydraulic history), but I generally draw upon evidence from England and northern Italy for the more detailed examples. This is not a comprehensive gazetteer of European water systems or components, nor do I claim to have included every possible variant. Instead, I have sought to identify the main problems that were encountered at each stage in the development and use of water systems and to indicate a range of possible strategies available for solving them.

I have not attempted to force my data into a rigid theoretical framework, and I have tried to limit my use of theoretical jargon. Nevertheless, I would like to acknowledge my intellectual debts to the innovation-diffusion processes defined by Everett Rogers and Floyd Shoemaker, the model for the stages in the evolution of large technological systems proposed by Thomas Hughes, and the emphasis on perceptions of technology by the proponents of the social construction of technology (SCOT).[5] These theoretical approaches, together with studies of the introduction of modern water systems in developing countries, have helped me frame questions and organize a jumble of individual fragments of information into a larger (and I hope more coherent) picture. The following chapters attempt to chart a course between the Scylla and Charybdis of technological and social determinism. While writing them I have tried to strike a balance between the technical and human aspects of medieval hydraulic systems and to remember that beneath the welter of documents and diffusion patterns, configurations and components, ordinances and expenditures, lie the perceptions, the choices, and often the plain hard work of individual men and women.

Ingratitude . . .
is a burning wind
which dries up
the fountain of piety,
the conduit of mercy,
the floods of grace.

SALIMBENE DE ADAM
(misquoting Bernard of Clairvaux)

Acknowledgments

It is a pleasure to have the opportunity to express my gratitude to the institutions and individuals who have assisted and encouraged me in the course of preparing this book. In particular I would like to acknowledge the courteous and expert assistance given to me by the staffs of the Bodleian Library, the Ashmolean Museum, the Biblioteca Comunale di Viterbo, the Canterbury Archaeological Trust, the Winchester Museums Service, and the University of Oklahoma library system. Robert Brentano, Gene Brucker, Ruth Tringham, and David Whitehouse supervised my Ph.D. dissertation, which forms the nucleus of this study. My senior colleagues in the History Department at the University of Oklahoma have been generous with their support and encouragement—and with their gentle and persistent nagging to get the book out!

During the course of this project I received financial assistance from an Ehrmann Fellowship, the Mabelle McLeod Lewis Memorial Fund, and the University of Oklahoma. I can only hope that the end results justify their generous support. Andrea Hood permitted me to share her apartment in Italy and introduced me to the Viterbo archives. I was able to base my research in Oxford for extended periods of time thanks to Alan Beggs, Elizabeth Morse, and Bob and Sue Ollenbuttel. While I was preparing the manuscript, Jamie Hart, Victoria Morse, and Bill North read individual chapters; Melissa Stockdale, Don Pisani, and Ann Kipp read over the full draft. Their thoughtful suggestions and comments have been invaluable and have saved me from some embarrassing missteps. Margaret Smith, Earl Chain, Jr., and Charlotte Jenkins from the University of Oklahoma Instructional Technology Services helped me prepare the illustrations. It has been my good fortune to work with a press that treats its authors with unfailing professionalism and courtesy. In particular I would like to thank Robert Brugger and Melody Herr for deftly guiding me through the stages of the publication process and Lois Crum for her careful copyediting.

Finally, I would like to thank my students and my family for their patience and support. My parents, Robert and Barbara Magnusson, taught me to love history, navigated some of the narrowest roads and steepest grades in Europe so I could visit remote sites, and were always ready to lend a helping hand. Sadly, my father did not live long enough to see the book finished. My nieces, Kelly and Amy Magnusson, cheerfully endured a summer camping in Oxford while their aunt conducted her research. My students' enthusiasm for medieval history keeps my own from flagging, and their perceptive questions challenge me to look at familiar sources from fresh perspectives. Last but by no means least, I would like to thank Kip for countless hours spent trawling through likely (and unlikely) sources, for reading through the manuscript in all its various incarnations, for rigorous critiques, and for keeping me fed, clothed, and pointed out the door in the right direction.

The errors and shortcomings that remain are entirely my own.

Water Technology in the Middle Ages

1 Survival and Revival

During the years 1220–22, an underground pipeline was laid by Master Laurence of Stratford to supply fresh water to the Augustinian canons at Waltham Abbey in Essex. Before suspending work for the first winter, Laurence was able to check the pressure in his partially completed system (and engage in a little showmanship) by erecting a vertical pipe twelve feet high at the point where work had temporarily ceased. Upon letting the water in, the pipe's perforated terminal "gave forth water copiously," to the admiration of a "great procession" of people who came to view it.[1]

The Waltham spectators were not alone in their enthusiastic response to the wonders of hydraulic engineering. In Siena in 1343, nobles and commoners, artisans, women, boys, clerics, and peasants all feasted, drank, danced, and sang together "without a hint of scandal" in the city's main piazza, the Campo. The occasion for these festivities, so magnificent that chronicler Agnolo di Tura del Grasso found it "incredible to write or tell" about them, was the first gush of water into a new civic fountain. The celebration was, admittedly, slightly marred by the unfortunate death of a workman, who had been incautiously standing in the subterranean channel and was drowned as the water rushed in. Nevertheless, the Sienese had good reason to rejoice: in spite of unforeseen delays and cost overruns, a fountain worthy of the honor of the city stood at last in its very heart.[2]

Although complex hydraulic systems were once again being built throughout western Europe in the High Middle Ages, the ability to harness and direct the flow of water was not yet common enough to be taken for granted. For the crowd at Waltham, the spectacle of water gushing forth from a pipe must have been an amazing novelty. Siena's celebrants were more hydraulically sophisticated—their commune had, after all, been providing public fountains for a century and a half—but the creation of a new fountain was still an occasion for a memorable public festival.

Were these new water systems of the High Middle Ages the product of the continuous survival of a Roman technological tradition or of its revival? Medieval gravity-flow water systems clearly owe a considerable debt to the engineering achievements of the classical world. Technologically similar water systems had already been widely and successfully deployed throughout the western Roman Empire. It was not necessary to make significant modifications to Roman-style physical components in order to adapt them to a new climate or a new topography. Medieval water systems were not simply composed of physical artifacts and natural resources, however. The networks of interconnected hydraulic components were inextricably linked to their local social environments. The process of technology transfer required the adaptation of Roman hydraulic engineering to the characteristics of a different time, not a different place—a different cultural geography, not a different physical one.

The basic technological trajectory of ancient hydraulic engineering, based on gravity flow and low-pressure systems of channels and pipes, was retained in the Middle Ages. The physical components of medieval water systems were based on a highly developed range of Roman models. Medieval pipes and taps are often so similar to their ancient counterparts that they can be misidentified as Roman if found out of context. What is less certain is whether this morphological similarity derives from an unbroken technological tradition or from a deliberate imitation of antique exemplars. Islamic hydraulic engineering, which was also derived from the technological traditions of antiquity, may also have played a role in the revival of water systems in Latin Christendom.

Whatever degree of continuity existed on the technological front, there was little continuity on the social side. The political, religious, and social upheavals of late antiquity and the early Middle Ages produced fundamental changes in social organization and cultural values, which triggered contingent changes in the social components of hydraulic systems. Many ancient systems collapsed altogether and were discarded; others underwent radical restructuring, as old physical components were reemployed to meet new goals.

The early medieval evidence presents a patchwork of continuity and discontinuity, discard and renewal. The old assumption that Roman aqueducts were either destroyed by barbarian invaders or robbed for their lead in late antiquity is being replaced by a more nuanced picture of local and regional differentiation. A growing body of evidence demonstrates that in the centuries following the so-called fall of Rome, at least some ancient aqueducts were restored or remained in use. In the city of Rome itself, popes Hadrian I (772–95), Paschal I (817–24), Gregory IV (827–44), Sergius II (844–47), and Nicholas I (858–67) undertook repairs to four of the classical aqueducts (the Aqua Virgo, the Traiana, the Claudia, and the Iovia). A Roman aqueduct at Brescia was reutilized from 761 onward to supply the convent of San Salvatore. Some Roman drains and fountains may also have continued to function. Saint Cuthbert saw what may have been a working Roman fountain on his visit to Carlisle in 685. The monumental Roman sewers in Pavia, familiar to Liutprand of Cremona in the tenth century, were still familiar to Opicino de Canistris in the fourteenth and appear to have flowed continuously throughout the Middle Ages.[3]

There were probably regional differences in the survival of hydraulic structures. In Britain, most Roman water systems do not seem to have outlasted the Roman occupation, but in northern France it has been estimated that about half of the Roman aqueducts were still functioning in the Merovingian period. A detailed assessment remains hampered by the uncertainty inherent in ascertaining the dates of abandonment for individual structures. The Pont du Gard, for example, was once thought to have gone out of use as a result of the Norse invasions of the ninth century, a date apparently supported by the encrustation of calcium carbonate (sinter) in the channel. Recent archaeological work, however, suggests an earlier date (perhaps the sixth century) for its abandonment, though some segments of the channel may have remained in use until considerably later.[4]

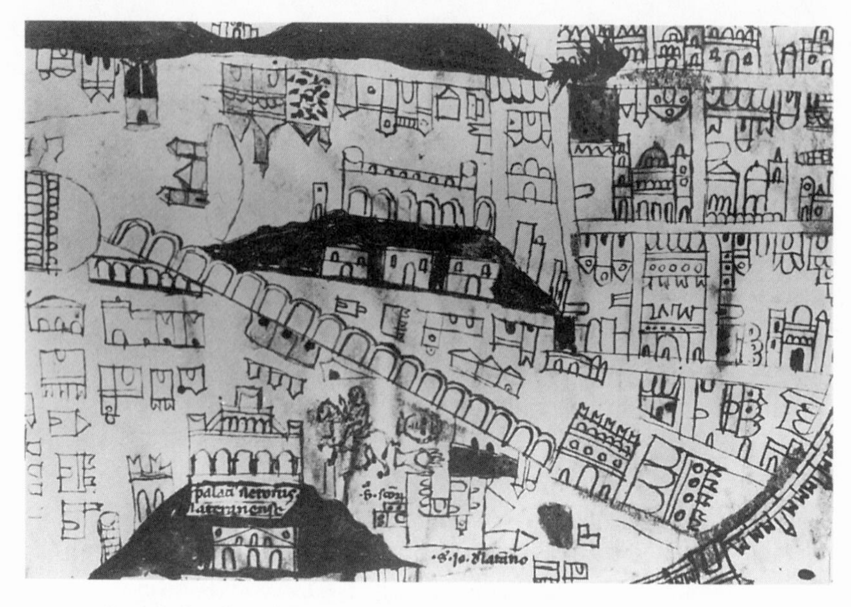

FIG. 1.1. Detail of a plan of Rome, showing the arcades of the Aqua Claudia. Some Roman aqueducts continued to flow under papal sponsorship in the early Middle Ages. Their arcades were a visible reminder of Roman hydraulic traditions. Paolino Veneto, "Chronologia magna" (c. 1323), Marciana Ms. lat. Z. 399 (1610), fol. 98. Reproduced in Bernhard Degenhart and Annegrit Schmitt, eds., *Corpus der Italienischen Zeichnungen, 1300–1450,* vol. 2, no. 3 (Berlin: Gebr. Mann Verlag, 1980), pl. 12.

The early Middle Ages also witnessed the construction of some entirely new water systems. Unlike their Roman predecessors, which were built primarily by secular sponsors to feed luxurious baths, early medieval waterworks were usually built by ecclesiastical patrons to supply baptismal fonts, atrium fountains, monasteries, and much more modest bathing establishments. Several early medieval palaces were also provided with complex water systems.[5]

In spite of this degree of continuity, however, the broader picture was one of technological discard and regression during late antiquity and the early Middle Ages. The majority of Roman water systems gradually decayed and were abandoned. Most communities obtained their water from rivers, wells, and cisterns, a simpler level of technology appropriate for a less urbanized society. The political turmoil and breakdown in authority precipitated by internal strife and a second wave of invasions during the ninth and tenth centuries may have triggered the final abandonment of some surviving systems. The severe social and economic disruptions of this period, which seem to have left some communities without the will or resources to maintain or rebuild their water systems, were probably more significant than any physical damages inflicted by hostile troops.[6]

By the late ninth and tenth centuries, the thread of technological continuity had become seriously frayed, but it does not seem to have completely snapped. The evidence for this period is sparse and difficult to assess. There are occasional references to new hydraulic projects. An attempt (c. 835) to cut an aqueduct through the mountains for the monastic mills at Lobbes, Belgium, had ended in failure. In spite of this setback, the monk Notjer designed a complicated new water system for Lobbes in 974. Louis II granted the bishop of Verona permission to take a water pipe across a public bridge in 873 and gave the monastery of San Sisto at Piacenza permission to restore the ancient aqueducts or build new ones in 874. The monastic complex at Saint Gall, Switzerland, seems to have had some sort of water system (perhaps wooden pipes?) in 890. The "aqueduct which runs below the dormitory" at Abingdon Abbey, built by Aethelwold (954–63), though, was probably a simple channel.[7]

It is uncertain how many earlier conduits continued to function during these unsettled centuries. The successors to pope Nicholas I (858–67) seem to have abandoned the tradition of papal repairs to aqueducts. A late-seventh-century conduit at Saint-Denis was sacrificed to build fortifications in 869. A Carolingian aqueduct at Farfa may have physically survived a devastating fire in 897, but there is no indication that it remained in operation in later years. The Fulda water system installed by Abbot Sturmi (744–99) is one possible survivor. By the mid–twelfth century the old system was seriously deficient (if functioning at all), but at least the monks were still aware of it. Abbot Marcuardus (1150–65) had it restored and put back in working order. Some southern Italian water systems, such as the aqueduct at Salerno, apparently did continue to operate during these centuries.[8]

Whether the morphological similarity between later medieval and Roman hydraulic components represents a survival or a revival of Roman techniques, the complex water systems that began to reappear in eleventh-century Europe were, for all practical purposes, innovations. The readoption of complex water systems in the High Middle Ages was a phenomenon that occurred throughout Europe. Construction of new complex hydraulic systems seems to have begun in the early eleventh century, and it became increasingly widespread during the twelfth and thirteenth. The concentration of early systems seems higher in Germany than elsewhere, although whether this reflects the actual medieval situation or is a product of particularly assiduous German scholarship remains to be seen. German palaces and monasteries with early piped systems include Goslar (1036), Essen (1039–58), Harzburg (c. 1065–74), Hirsau (1092–1105), Grosskomburg (c. 1100), Bamberg (c. 1117–39), Prüfening (1121–63), Magdeburg (1125–60), Erfurt (1136), Ensdorf (twelfth century), and Goseck (twelfth century). Several monastic houses in Lotharingia also had complex water systems by the twelfth century.[9]

Other early systems include Chartres (c. 1090), Saint Bertin (1095–1123), Paris (before 1119), and Salzburg (1136). Cluny had some sort of subterranean conduit delivering water to the various monastic offices by the time of Peter Damian's visit in 1063. According to Orderic Vitalis, a subterranean conduit supplying the citadel at Alençon was cut during the siege of 1118. Saint Bernard's reluctance to leave the site of the first Clairvaux in 1133 was based, in part, on the fact that a costly water system had recently been built there. The second Clairvaux was also supplied with buried conduits to various buildings. In Rome, Calixtus II (1119–24) revived the tradition of papal hydraulic sponsorship by restoring water to the city. He also built the Mariana aqueduct, which utilized part of the old Aqua Claudia channel and collected waters from the ancient Aqua Iulia and Aqua Tepula to provide water for the Lateran. Aqueducts fed royal palace fountains in twelfth-century Sicily. The earliest known piped system in Britain comes from Wolvesey Palace in Winchester (c. 1129– 35). Other twelfth-century British systems include Canterbury (1153–67), Lichfield (by 1166?), Kirkstall (c. 1160–82), Evesham (1160–89), Gloucester (1163–84), Westminster Palace (1169–70), Churchdown (1170–81), Fountains (1170s), and Waverley (1179). Most of these complex water systems of the eleventh and twelfth centuries were associated with palaces or monasteries. There were a few early urban fountains (two Sienese fountains are mentioned in a document of 1081, one of Bergamo's fountains was in existence in 1156, and Viterbo had at least one by 1192), but most urban systems were products of the thirteenth century or later. Some castles, palaces, manor houses, hospitals, and gardens were also provided with piped water and fountains in the High Middle Ages.[10]

The social and geographical limits of the diffusion process have yet to be fully defined. In the north, several Scandinavian monasteries built complex water systems, but as yet there is no evidence for piped water in medieval Swedish towns.[11] On the eastern frontier, the inquiry has come to include Poland, Slovakia, and Hungary. At the peripheries of Western Christendom, more complex diffusional crosscurrents may have come into play, with Byzantine influences to the east and Islamic hydraulic traditions along the eastern and southern shores of the Mediterranean, in Sicily, and in the Iberian peninsula.

There were a number of aspects of the new social environment of the High Middle Ages that proved to be favorable for the readoption and diffusion of complex water systems. It was a period of extraordinary monastic expansion, astonishing even to contemporaries. "Like a great lake whose waters pour out through a thousand streams, gathering impetus from their rapids, the new monks went forth from Cîteaux to people the West." Hydraulic technology was carried along on the flood, part of a cluster of innovations affiliated with the architectural expression of an institutionalized way of life. At the practical level, prevailing religious attitudes facilitated monastic access to natural and financial resources. Laymen donated springs, rights-of-way, and sometimes money for monastic conduits as expressions of piety and in the expectation of spiritual rewards. More intangibly, the new "optimistic rationalism" that emerged in the theology of the twelfth century created a religious atmosphere favoring technological undertakings. Although water retained its earlier sacramental symbolism and holy associations, it was also seen as an element in a natural world that was (at least potentially) intelligible and predictable. God's ordered creation could be understood and mastered by man, and man's own creative activities took on religious significance through their relationship to God's work.[12]

The renewed growth of cities created an increased demand for clean urban water. At the same time, the establishment and growing political clout of municipal governments provided administrative organizations that were interested in, and capable of delivering, improved urban infrastructures. Once civic administrations were able to promulgate and enforce their own statutes, keep records, raise money, and hire public employees, the organizational tools to sponsor, finance, administer, and regulate large-scale water projects were in place.

Contemporary developments in the construction industry also favored the readoption of hydraulic technology in the High Middle Ages. The increased demand for stone buildings meant that more trained masons were available to turn their hand to stone-lined water channels and drains. Plumbers, already experts in casting sheets of roofing lead, had only to learn how to bend and join their sheets to make pipes. The revival of wheel-thrown pottery made it easier for potters to manufacture earthenware pipes. The expansion of the product lines of some kilns to include other types of building components may also have served as a stimulus to pipe manufacture, by bringing potters into contact with members of the building trades. Active quarries, a lime industry, a lead industry, and transportation networks were all in place to provide and deliver materials.

Other factors, however, worked against the spread of hydraulic technology. Springs, the preferred source for intake systems, were not universally available, whether because of local geology or uncooperative landowners. The nature of gravity-flow systems constrained the selection of a conduit route between the water source and the destination within relatively fixed topographic limits: access to land for conveyance systems could be denied if it required crossing the property of a hostile landowner, breaching a city wall, or digging up public streets. Laws could be used to promote and protect hydraulic systems, but they could also be used to obstruct them.[13] Potential sponsors often had other priorities, and many still lacked the organizational and financial resources to support the construction and operation of a complex water system.

Knowledge about hydraulic technology almost certainly spread along established communications networks. This took place at two levels: the spread of nonspecialist knowledge about water systems to potential sponsors, and the transmission of detailed technical knowledge to craftsmen. The study of medieval communications channels has been hampered by the tendency of scholars to restrict themselves to the study of a single region or type of institution. This involves an implicit assumption that the meaningful links took place between like individuals or institutions. Modern diffusion research, however, suggests that communication across social boundaries also plays a crucial role in the diffusion process, since close associates and near peers are less likely to have unfamiliar information than those who belong to different social or regional groups. Both internal and cross-boundary diffusion mechanisms need to be better defined for medieval hydraulic technology, but a few preliminary generalizations are possible.[14]

The spread of nonspecialist knowledge about water systems would have been relatively easy: although conveyance networks were hidden, distribution structures were highly visible, and their potential benefits would have been apparent to even the most casual observer. The location of civic fountains in main squares and streets meant that they would have received maximum public exposure. It was not necessary for a potential adopter to understand hydraulic principles to appreciate the beauty and utility of a fountain or to figure out how to use it.

A good deal of the initial general awareness of hydraulic technology must have been the result of the widespread physical mobility that characterized the High Middle Ages. Merchants, pilgrims, travelers to the papal court, and university students would all have come into contact with complex water systems, whether in monastic guest houses or in cities. Travelers going to Rome on the Via Francigena passed through Siena, whose fountains were famous. Once at Rome they could marvel at the ancient aqueducts and admire the Cantharus fountain in the atrium of Saint Peter's, even when the water no longer flowed. During much of the thirteenth century, the papal court was to be found outside of Rome; it was often in Viterbo, a city with prominent public fountains. For the later thirteenth century, the same could be said for Orvieto and Perugia. The international body of students at Paris lived in a city that had had a public water supply since the late twelfth century. Merchants bringing their wares to London or Bristol had the opportunity to see numerous public conduits. Sailors embarking from Bristol, such as the crew who conveyed Margery Kempe on her pilgrimage to Santiago de Compostela, could provision their ships with fresh water from a conduit built right on the quay. Having reached Santiago, Margery and her fellow pilgrims would have been supplied with water from a fountain constructed especially for them at the side of the cathedral. Pilgrims to the shrine of Saint Thomas at Canterbury enjoyed the use of a fountain-laver in the monastic guest house; visitors to Kirkstall Abbey were served by a piped water system to the guest quarters. Western Crusaders passed through, conquered, and settled in eastern cities such as

FIG. 1.2. Extramural urban fountain. The beauty and high visibility of fountains helped spread awareness of hydraulic technology. Titus Livius, Mailand, Ambrosiana MS C. 214 inf., fol. 75v., 1372–73. Reproduced in Bernhard Degenhart and Annegrit Schmitt, eds., *Corpus der Italienischen Zeichnungen, 1300–1450*, vol. 2, no. 3 (Berlin: Gebr. Mann Verlag, 1980), pl. 47.

Antioch, Caesarea, and Constantinople, which had highly sophisticated water systems. Such casual contacts were undoubtedly instrumental in spreading a general awareness of hydraulic technology, but they may not have been sufficient to directly trigger subsequent adoptions. Nor did they necessarily convey much in the way of technical know-how: the French author of a twelfth-century guidebook for Santiago pilgrims described the cathedral fountain with great enthusiasm, but he had no idea where the water came from or where it went.[15]

Modern research on the diffusion of innovations indicates that potential adopters are most likely to be persuaded to take the plunge when they receive evaluations of the innovation from people like themselves, whose subjective experiences with the innovation have been positive. Some medieval adoption decisions seem to fit this pattern and were probably the result of direct personal ties. Henry de Blois, bishop of Winchester (1129–71), the probable patron of the lead-pipe system at Wolvesey Palace, had been a monk at Cluny, which had enjoyed a water system since at least the mid–eleventh century. At the time Lichfield Cathedral's water system was built (shortly before 1166?), the bishopric of Lichfield was held by Walter Durdent, a former prior of Christ Church, Canterbury. His subprior and eventual successor at his old house had been Wibert, the sponsor of the Christ Church water system, built between 1153 and 1167. An eleventh-century aqueduct at Hirsau was financed by a layman, Wignand, a citizen of Mainz, who later became a monk at Grosskomburg. It seems probable that Wignand's personal knowledge of the Hirsau system played a part in the decision to build the late-eleventh-century Grosskomburg waterworks. The establishment of Dublin's water system in 1244 may have been due in part to the many settlers from Bristol, a city that already by that date had conduits serving Saint Mary Redcliffe, Saint John's hospital, Saint Mark's hospital, and the Dominican friary.[16]

Monastic hydraulic diffusion was not confined to any one order: complex water systems are known from nearly all types of religious houses, including nunneries. A few exceptional monasteries may have acted as "opinion leaders." The enormous spiritual prestige of a Cluny, a Hirsau, or a Clairvaux may have stimulated lesser houses to imitate their water systems as well as their observances and architecture. The rapid and widespread diffusion of the Cistercian order was followed by a rapid and widespread diffusion of hydraulic technology, in a kind of institutional cloning.

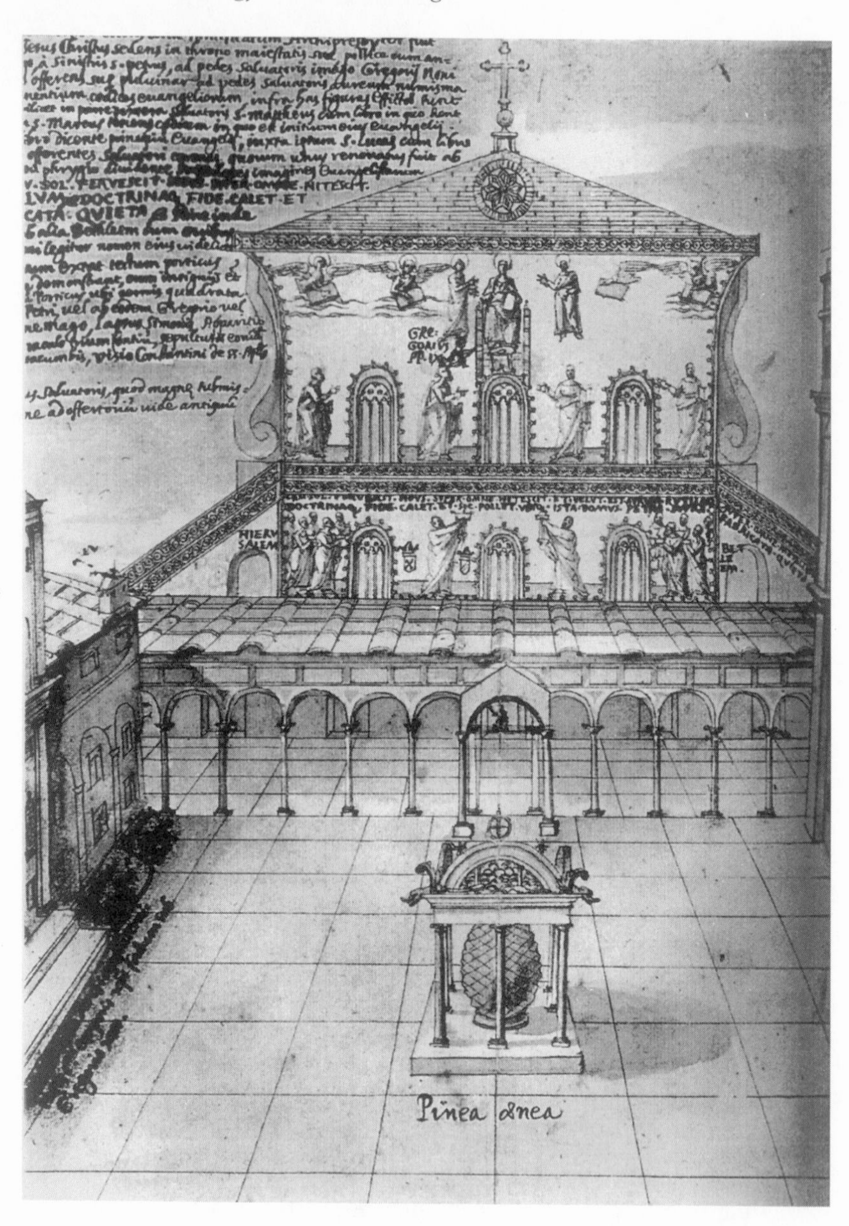

FIG. 1.3. Cantharus fountain. Medieval pilgrims to Rome would pass by the antique atrium fountain as they entered Saint Peter's basilica. Watercolor by D. Tasselli (c. 1611). Biblioteca Apostolica Vaticana, Archivio di S. Pietro, Album A 64 ter, fol. 10. Reproduced in Angiola Maria Romanini, ed., *Roma nel Duecento: L'arte nella città dei papi da Innocenzo III a Bonifacio VIII* (Torino: Edizioni SEAT, 1991), 274.

The annual General Chapter meetings and visitations were mechanisms by which architectural directives were transmitted throughout the order and probably served as ongoing communication channels for information about hydraulic engineering. Filiation links between houses may also have played a role. The Cistercians seem to have had something of a policy of sending out architects to help instruct the members of new foundations. In 1133 Saint Bernard sent Geoffrey d'Ainai, a senior monk at Clairvaux, to instruct the new Cistercian community at Fountains. Since Geoffrey was himself a skilled architect, it is thought that he trained some of the Fountains monks in architecture as well as Cistercian customs. It is possible that he served as a source of information about plumbing as well as customs and liturgy. He would certainly have known Bernard's brother Gerard (also at Clairvaux), whom Bernard eulogized for his constant standard of excellence in matters great and small. "Did anything ever escape the skilled eye of Gerard in the buildings, in the fields, in gardening, in the water systems?" Although men such as Geoffrey and Gerard undoubtedly played important supervisory roles in Cistercian building projects, the professional craftsmen who became lay brothers rather than choir monks may have played equally important (if often unrecorded) roles in the transmission of technical expertise within the order.[17]

It is possible to compile some rough statistics for the monastic adoption of water systems in Britain. During the peak centuries of medieval monasticism, there were about one thousand religious houses in England and Wales. If one combines houses with documentary evidence for conduits (mainly conduit charters and licenses) with those for which there is archaeological evidence for a complex intake system (water pipes or a sophisticated *lavatorium* [washing place]), there appear to be some 130 houses that probably had advanced intake systems.[18] Obviously, these statistics should not be pressed too far: not every house that obtained a conduit license necessarily built a water system, not all of the archaeological evidence is conclusive, and there are almost certainly other sites with water systems that have not yet been identified. Nonetheless, if we take these figures as rough estimates, we find that about 13 percent of religious houses may have had some sort of advanced intake system. Using the incomes recorded at the Dissolution as at least an approximate guide to the relative wealth of the houses that adopted water systems, it becomes immediately apparent that income was a key factor in determining which houses could undertake water projects. Of the 130 houses identified as probable adopters, only a small handful had incomes of less than £100 a year (and of these six, two had obtained branch lines from other houses and three were questionable candidates for inclusion in the list). A financial threshold of £100 per annum would have put a complex water system beyond the means of about half of the religious houses. For houses above this limit, however, the rate of adoption was approximately 25 percent. The size of the community seems to have been less important than its income. Over 40 percent of the probable systems were associated with small houses (25 or fewer), and about 70 percent of the sponsoring communities probably had fewer than fifty members (not counting servants or lay brethren).[19]

The Benedictines had the largest absolute number of probable systems at 32 (25% of the total), followed by the Franciscans (26), the Augustinian Canons (22), the Cistercians (16), and the Dominicans (15). Orders with a few systems (1–5) included the Carthusians, the Carmelites, the Premonstratensians, the Austin Friars, the Gilbertines, the Bonhommes, and the Knights Hospitallers. If one looks at the rate of adoption by comparing the number of probable water systems with the total number of houses for each order, however, the ranking differs. The orders with the highest rates of adoption were the Carthusians (50%), the Franciscans (39%), and the Dominicans (26%).[20] Other orders fell well behind this rate: the Cistercians (12%); the Carmelites (10%); and the Benedictines, the Augustinian Canons, and the Premonstratensians (8% each). The rather average rankings of the Cistercians on both lists may indicate that the order's reputation as a leader in hydraulic technology has been somewhat overrated, whereas the high rankings of the Franciscans suggests that they may have played an important role in hydraulic diffusion.

Medieval convents have left fewer surviving records than male monasteries, and fewer have been excavated, so proportionally less evidence is available for assessing the degree to which women, too, adopted water systems. As potential sponsors of hydraulic systems, nuns stood at a considerable disadvantage compared to their male counterparts: they commanded fewer financial resources, had greater difficulty acquiring land beyond their initial endowments, and were less likely to engage in ambitious drainage and land reclamation projects. Moreover, they could not necessarily count on the full backing of men, even within their own orders. Cistercian monks, for example, were very reluctant even to acknowledge

FIG. 1.4. Carmelite friar dipping a jug into a fountain. The two glass beakers sitting on the fountain rim suggest that the water will be used for drinking. Although Cistercian waterworks are the best known, many medieval monastic orders built water systems. Pietro Lorenzetti, *Carmelitani al pozzo* (detail). Reproduced in Amerigo Restucci, ed., *L'architettura civile in Toscana: Il Medioevo* (Siena: Silvana Editoriale, 1995), 499, fig. 1.

women's houses, and it would seem that women were often left out of the loop when it came to monastic communication channels.[21]

In spite of these impediments, nuns were early and important sponsors of gravity-flow water systems. The lead pipes from Abbess Theophanu's convent in Essen date to the first half of the eleventh century, which makes them the earliest surviving example of medieval lead pipes in Germany. They probably fed an atrium fountain and then ran under the church to supply the cloister.[22] In England, the nuns at Godstow had a conduit that may have been built as early as 1135 (which would make it among the very first medieval systems in Britain). Although a later twelfth-century date is also possible, this would still have been quite early by British standards.[23] One of the best-studied Cistercian piped systems belonged to the nuns at Maubuisson; it was under construction within two years of the founding of the convent in 1236. Essen, Godstow, and Maubuisson were exceptional, inasmuch as all enjoyed royal connections or royal patronage. Abbess Theophanu was an Ottonian princess, the niece of Otto III, whereas Maubuisson was founded by Blanche of Castile, queen of France. Godstow, although not a royal foundation, received generous support from both King Stephen and Henry II (who became the abbey's patron following the death of his beloved mistress, Rosamund Clifford, who was buried there), as well as the local nobility. In England alone, however, other medieval water pipes are known from Augustinian convents for women at Lacock, Clerkenwell, and Grace Dieu; the Benedictine Nunnaminster at Winchester; and the Franciscan Minories house in London; at Wherwell Abbess Euphemia (1226–57) "with maternal piety and careful forethought" constructed a stream-flushed great drain. Although evidence for conduits has been identified for only a little over 5 percent of English nunneries, for the small number of houses with incomes in the range of £100 or more per annum, the estimated rate of adoption (22%) is very similar to that for male houses (26%). In spite of the obstacles they faced, at least some women religious seem to have been quite up-to-date in their technological awareness. The potential advantages of a water system clearly appealed to women as well as men, and the communities of women that could command enough resources to construct one do not seem to have lagged significantly behind their male peers in their adoption of the technology.[24]

There are occasional hints about the spread of hydraulic knowledge between ecclesiastical orders. In his memoirs, written in a small Benedictine abbey near Laon in the early twelfth century, Abbot Guibert of Nogent described the water system at La Chartreuse in some detail: "They have water for both drinking and other purposes from a conduit, which goes around all their cells and flows into each through interior holes in the walls." Guibert's source of information may have been one of the monks from Nogent, who had been the companion of Bishop Godfrey of Amiens when he was in exile at the Carthusian house in 1115. Guibert was clearly impressed, but there is no indication that the news from La Chartreuse motivated him to attempt to supply his own convent with similar amenities—though perhaps he helped inspire an interest in hydraulics among some of his readers. A more direct case for the transmission of technical knowledge comes from Denmark. In 1175 the Victorine canons at Aebelholt were assisted in the construction of their conduit by Brother Stephen, who belonged to the Cistercian abbey at Esrum. Brother Stephen's hydraulic expertise proved to be so valuable that Abbot William wrote a letter begging the abbot of Esrum to allow Stephen to stay on for a few more days until the conduit was finished. In the mid–thirteenth century, a canon of Hexham wrote a letter of introduction for another Stephen to the cellarer of Tynemouth Priory: "I am sending you Stephen de Len, who is an honest workman, and, as I have heard, is skilled in plumbing and in laying on water. Do not think the worse of him for his shabby clothes. He has two or three times lost his all in this war [the Barons' War], which is hardly yet over."[25]

Unlike the church or dormitory, a sophisticated water system was not an essential primary component of a monastic complex, however. New houses were seldom able to construct a complex system immediately, even though the availability of water resources often played an important role in their choice of a site. Monasteries were most likely to build a conduit after the house had become firmly established, with well-developed local networks of power and patronage and sufficient financial resources to tackle major building programs. At Norton Priory, archaeological excavations have revealed the development of both a primitive water-management system, associated with the priory's first, temporary timber structures, and the more sophisticated, permanent replacement of that system. The original twelfth-century system consisted of a network of open drainage ditches and gullies, a large drain lined with unmortared and roughly shaped stones, and a hollowed tree-trunk drain below the latrine block. By the end of the century, the expanding community had mustered sufficient resources to begin the construction of permanent stone buildings. An important element in the building program was the installation of a vastly improved water system, which included spring-fed lead intake pipes supplying basins and cisterns, stone-lined drainage channels, and a masonry great drain flushed by water from the moat.[26]

City-to-city links were also channels for the transmission of technological awareness. Foreign visitors to Siena and Viterbo not only marveled at the beautiful fountains but must also have carried at least a general impression concerning these wonders back to their own communities. More prolonged residence in a city with a civic water system would increase the chance of gaining firsthand knowledge. Italian communes, for example, hired outsiders to fill the office of podesta and often sent their own leading citizens into exile; furthermore, Italian merchants frequently took up semi-permanent residence in foreign markets. Such men would have had the chance to become familiar with the physical components of water systems and also to learn about their administrative requirements. Since they were likely to have close ties to the urban governing elite and to hold municipal offices themselves, they would have been in a strong position to influence their own city's water policy on their return home.

At the local level, the successful adoption of a water system by one subgroup could lead to secondary diffusion across social boundaries. Though most cities had no complex systems at all, those that did have them often ended up with several monastic conduits, multiple municipal conduits, and networks of branch lines to other institutions, such as hospitals and churches. In Bristol the hospitals of Saint Mark, Saint John, and possibly also Saint Bartholomew were probably fed by pipes from the Greyfriars' supply. The London Charterhouse springs also fed a pipeline to the nuns at Saint Mary Clerkenwell. In Smithfield the hospital of Saint Bartholomew leased a supply of piped water from the nearby priory.[27]

Religious houses functioned as vital communication channels in the diffusion of technological awareness to medieval townsmen. In many cities the first water systems belonged to an urban friary or cathedral chapter. These served as technological exemplars and in some cases became the actual nuclei of civic systems. The Christ Church, Canterbury, waterworks plan portrays a fountain outside the main precinct, which would have been accessible to the general public. Even cloisters were not necessarily inaccessible to laymen (or, more scandalously, women), as the visitation records of exasperated bishops repeatedly show. According to the records of Bishop Alnwick of Lincoln (1436–49), Daventry Priory was allowing laywomen to come through the cloister to draw water from the monastic laver, to the great scandal of the house: "Women have general resort to the kitchen and to the washing places [*lavatoria*] in the cloister, where they get up on the edge to fill their pots at the washing places, and so they befoul the same edge with their feet." Alnwick also received the complaints of the nuns of Godstow Abbey, where secular serving-folk and other laywomen were availing themselves of the claustral latrine (though perhaps not the cloister conduit, since it was in urgent need of repairs at this time). The sisters asked that the laywomen be forbidden access to the convent's facilities, but with a nod to Christian charity, they also requested that "another place be appointed them to this end *outside* the cloister." In Lichfield, the dean complained to Bishop Geoffrey Blythe that women fetching water from the conduit in the Cathedral Close were causing no small scandal to the inhabitants. Such opportunities for the laity to closely observe and to use ecclesiastical waterworks, even if distressing to punctilious bishops, allowed secular citizens the chance to give the technology a trial run before they committed themselves to the construction of a municipal system.[28]

The detailed mechanisms for the transmission of specialist hydraulic knowledge in the Middle Ages remain rather obscure. It is possible that ancient literary texts played some role. Vitruvius's *De Architectura* would have been available in a number of monastic libraries. The other most potentially useful Roman text Frontinus's *De Aquis Urbis Romae,* was, however, apparently virtually unknown in the Middle Ages, although one copy was kept at Monte Cassino. It is difficult to demonstrate whether or not ancient texts exerted a direct influence on medieval hydraulic technology. It is possible that they may have inspired the occasional monk to turn his hand to engineering, but one suspects that they would have been of value mainly to the reader who already possessed some related technical knowledge.[29]

The intensification of urban construction in the High Middle Ages, along with the deeper foundation trenches required for stone buildings, may have helped to spread awareness of hydraulic technology among construction workers, by bringing greater numbers of Roman pipes and channels to light. The ability of medieval plumbers, masons, and other craftsmen to understand and imitate the products of their ancient counterparts may account for the apparent continuity in the design of hydraulic components such as pipes, taps, and masonry channels. The common practice of recycling ancient materials would have meant that rediscovered hydraulic components tended to find their way into the hands of the craftsmen most likely to understand them. Metal components were probably sold as scrap metal: lead pipes to plumbers, bronze taps to bronze smiths, etc. It is known that the masonry of some Roman aqueducts, and even the calcium carbonate deposits (sinter) were quarried for reuse. Masons involved in this *spolia* trade would, incidentally, gain a firsthand familiarity with the way Roman water channels were constructed.[30]

Because of the itinerant nature of the building trades, it was not necessary to have a hydraulic engineer living locally: outside experts could be brought in, though finding and recruiting them might require persistence. Master William the conduit maker (*conductarius*), who installed a new water system in Westminster Palace for Henry III, came from Reading—the king ordered that his traveling expenses be paid. Master Laurence, the man in charge of building Waltham Abbey's conduit, was based in Stratford. The engineer Maurice was responsible for twelfth-century lead-pipe distribution systems at both Dover Castle and Newcastle. The hunt for capable masters to build the Perugia aqueduct and fountains preoccupied the city's Consiglio Generale for years. In their quest for fountain-masters, they dispatched envoys to likely Italian towns, such as Viterbo, and inquired among the friaries. They finally found and hired an engineer, Bonomo of Orte, but before much progress had been made, Bonomo died, and the council had to renew their search. Eventually they managed to recruit Boninsegna of Venice (who had been discovered working on a fountain in Orvieto), Brother Leonardo of Spoleto, Brother Alberic (a Franciscan friar), Master Guido of Città di Castello, Master Coppo of Florence, and Brother Bevignate (a Benedictine monk).[31]

Large building projects probably played a key role in the diffusion of technical expertise among craftsmen, whereas the employment of laymen in the construction of cathedrals and monastic buildings served as a communication bridge between lay and ecclesiastical subgroups. The mix of friars, monks, and laymen at Perugia reflects a more general phenomenon. Some hydraulic experts were members of religious communities; some were lay craftsmen. The latter do not seem to have been restricted to any one craft, though most were involved in either metalworking crafts or construction trades: goldsmiths, plumbers, carpenters, and masons. The papal fountains at Sorgues and Avignon were made by Jean Belhomme, an Avignon goldsmith and perhaps clock maker. By the fifteenth century, some French plumbers, such as Parisian Jehan de Foing, were also specialist *fontainiers*.[32]

Any hydraulic project was likely to require at least the temporary employment of numerous craftsmen and laborers. Sienese civic documents preserve records of tools, materials, and payments to the masters and workmen who dug the *bottini* (subterranean filtration conduits). Workmen (and women) were hired by the day, but the masters tended to be long-term specialists who worked on one particular *bottino*. On occasion men from nearby mining towns were employed—presumably their special expertise was advantageous when difficult strata were encountered. The account rolls for Exeter Cathedral record payments to laborers, plumbers and their assistants, masons, sawyers, carpenters, and servants, all of whom were engaged in work on the conduit. It seems likely that most hydraulic experts would have acquired their technical knowledge by working as apprentices or secondary craftsmen on such projects. In some cases, hydraulic expertise seems to have been handed on from father to son. At Waltham Abbey, Master Laurence was assisted by his two sons, Ralph and William. When Laurence unexpectedly died, just as the project was nearing completion, his sons may have finished it off.[33]

Decisions to adopt innovations are based on the recognition of needs, awareness of an innovation's existence, and a favorable attitude toward the innovation because of its perceived advantages. Although the diffusion of hydraulic knowledge was a necessary precondition for the adoption of hydraulic technology, a medieval community's technological awareness did not automatically result in a decision to build a complex water system. Indeed, adoptions of complex water systems remained limited in this period, a fact that has helped perpetuate the myth that medieval water supplies were invariably primitive and unclean. In order to assess medieval adoption decisions, then, it is necessary to consider potential sponsors' perceptions of their hydraulic needs, together with their perceptions of the advantages and disadvantages of hydraulic technology.

Religious houses seem to have been more willing than other potential sponsors to construct complex water systems. This owes something to the institutionalized nature of monastic life. The regulation and synchronization of the activities of multiple users meant that water supplies were required to meet exceptionally high peak demands. Monastic customaries define the particular points in the day when monks washed their faces and hands at the *lavatorium*. Shaving and bathing were not daily activities, but the whole community was expected to perform these activities at set times. The exceptionally large size of many monastic reredorters suggests that the latrines were also subject to high peak demands. A medieval monastery would have been comparable to a modern institution such as a school or a factory, where peak use occurs at breaks between lessons or shifts and at mealtimes. The requirements of such a closely regulated, synchronized communal life did not absolutely require complex water systems, but those requirements did help create a preexisting need that favored their adoption. A water system that could accommodate the simultaneous needs of large numbers of users was precisely suited to the monastic way of life. A community like Christ Church, Canterbury, trying conscientiously to observe the constitutions of a Lanfranc, would almost certainly welcome the water system of a Wibert.[34]

In addition to meeting the practical needs of a monastic community, a supply of pure water had ritual significance. When Edward I sought advice on giving alms to the Friars Minor, provincial minister Robert de Cruce advised him: "Our brethren at Oxford suffer grievously from the want of an aqueduct. For the water of the well which we draw daily, and which we mix every day with our food, and sometimes drink on penitential days, and (what is a more serious consideration) which we mix with the wine of the Sacrament is very corrupt. If, therefore, it should please your lordship to assign the said alms to make good these defects, I do not think a use more pleasing to God could be found."[35]

Why did medieval cities, with their less closely regulated social organization, choose to adopt complex water systems? Some of the factors defined in studies of present-day technological adoption also seem to have played a part in medieval decisions to build civic water systems. Richard Feachem, in an attempt to prescribe explicit goals for water projects in modern low-income communities, defines improvements in water quality, quantity, availability, and reliability as desirable aims. The most immediate benefits derived from these improvements in the water supply are savings in the time and energy formerly spent in fetching water and improved

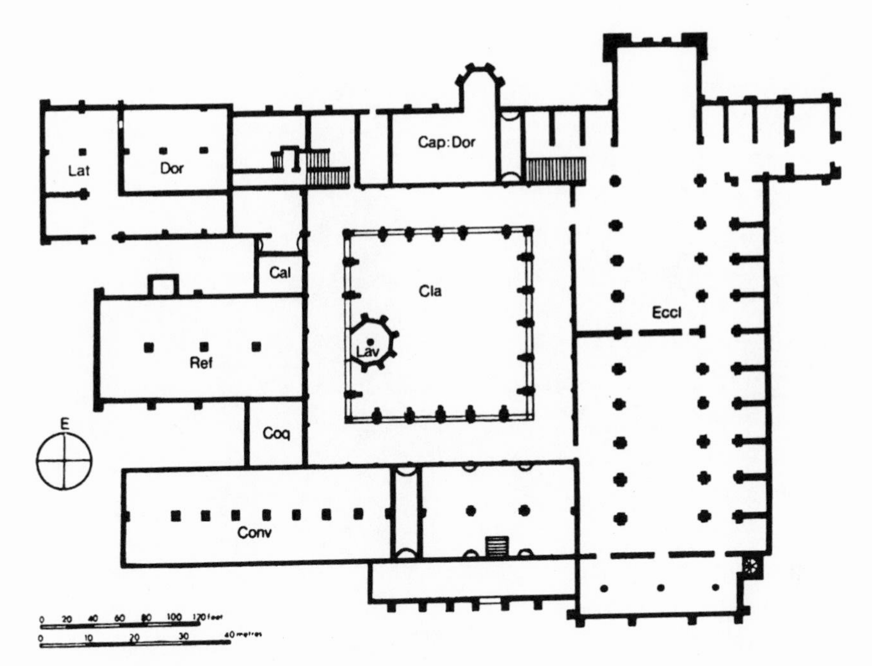

FIG. 1.5. Plan of the Cistercian abbey at Maulbronn. It was common to have the *lavatorium (Lav)* situated in the cloister near the refectory *(Ref)*, so that the monks could wash their hands as they entered. Here the fountain is enclosed in an octagonal fountain house. Christopher Brooke, *Monasteries of the World: The Rise and Development of the Monastic Tradition* (New York: Crescent Books, 1982), fig. 13.

health from eliminating or reducing the incidence of water-related diseases. George Foster, in an anthropological examination of the motivations that actually do underlie adoptions of technological innovations in traditional societies, isolates two motivations as "standing out far above all others": the desire for economic gain and the desire for high status and prestige.[36]

Surviving medieval documents seldom provide a comprehensive record of the decision-making process, so it is not always possible to determine which attributes were thought to be most advantageous. Such evidence as there is shows that improvements in water quantity and quality, convenience, and fire protection were considered practical advantages associated with complex hydraulic systems. More intangibly, beautiful fountains brought honor and prestige to their cities—hydraulic systems were not

FIG. 1.6. The monastic *lavatorium* at Fontenay, after a drawing by Viollet-le-Duc. The fountain's many spouts would have permitted a number of monks to wash simultaneously. Eugène-Emmanuel Viollet-le-Duc, *Dictionnaire raisonné de l'architecture française du XIe au XVIe siècle* (Paris: Bibliothèque de l'Image, 1858), 6:173.

merely functional components in the urban infrastructure but also served as expressions of civic pride and identity.

The adoption of complex water systems by medieval cities seems to have been stimulated by the rapid growth of urban populations and the consequent strains this placed on existing water sources. The increase in domestic and industrial consumption was coupled with an increase in the production of urban wastes: not only was the quantity of water from traditional sources often inadequate, but the increased levels of pollution meant that its quality (even by medieval standards) was unacceptable. Municipal statutes repeatedly addressed the related problems of water pollution and waste disposal; a few municipal governments looked to a technological solution.

Problems with the quantity and reliability of traditional water sources became more acute as settlements grew. Population estimates for medieval cities are notoriously unreliable, but there was a rapid demographic expansion during the twelfth and thirteenth centuries, followed by an overall population decline (sharply accelerated by the Black Death) in the fourteenth. The initial adoption of complex hydraulic systems by medieval cities correlates with the period of rapid growth: Paris, and perhaps Viterbo and Siena, instituted public systems in the 1190s, London acquired springs for its conduit in 1237, Dublin obtained permission to build a water system in 1244, and Perugia was contemplating an aqueduct by 1254. This chronological correlation is not coincidental: the growing urban demand for water threatened to overwhelm traditional sources of supply, and the increased production of urban wastes in the densely settled cities jeopardized the purity of both ground and river water.

The growth and decline of complex urban water systems were not simply functions of demographic cycles, however. London's population may have doubled in the thirteenth century and seems to have peaked around 1300: a recent analysis of settlement density in Cheapside suggests that the conventional estimate of 30,000–50,000 inhabitants should be revised upward. The initial construction of the Great Conduit in the mid–thirteenth century may well have been a response to the city's rapid demographic growth, but the expansion of the system in the late fourteenth and fifteenth centuries occurred during a period of prolonged demographic decline. Cities that adopted complex systems did not fit a single demographic profile: a few, such as London and Siena, were very large by medieval standards. Others, such as Bristol, were moderate-sized cities, and some, such as Exeter, were modest provincial towns. The majority of medieval cities, although equivalent in size to adopters of public water systems, did not sponsor hydraulic projects at all. Demographic pressure was a precondition that made the construction of an urban water system an attractive option, but it does not provide the sole explanation for the pattern of urban adoption.[37]

Hilltop settlements were particularly vulnerable to water shortages as their populations expanded. By the early thirteenth century, the town wells at Old Sarum were failing to satisfy the needs of the inhabitants. According to a litany of complaints about the chalky hilltop sent by the cathedral chapter to the pope, water had to be brought in from a distance "at a price that, elsewhere, would buy enough for the whole district." An inquiry into the chapter's grievances by a papal legate substantiated the charges, and in 1218 Honorius III issued a license permitting the chapter to transfer the site of their church to a more fitting place. In 1223 Bishop Richard Poore founded an entirely new town on a marshy site on the River Avon, which was amply supplied by a network of artificial water channels flowing down the centers of the streets and shallow wells sunk in the alluvial or gravel subsoils. New Sarum, or Salisbury, freed from the hydraulic restrictions of the old site, grew to be one of the largest provincial towns in England.[38]

Medieval Italian hilltop cities faced similar water shortages; rather than changing their sites, Orvieto, Perugia, and Spoleto adopted the less drastic expedient of building aqueducts to supplement existing supplies. Siena's adoption of an ambitious hydraulic system was probably a direct response to a serious resource deficiency. Wells and cisterns were (and remained) important sources of water, but their supply was insufficient to meet the demographic and industrial demands of the expanding medieval city. The arguments advanced in Siena's original decision to sponsor a public water supply have not been preserved, but later extensions to the system were frequently justified on the grounds that the existing *bottini* supplied insufficient water. The subterranean network was extended to follow every available vein, until more than twenty-five kilometers of bottini supplied water to Siena's civic fountains. As members of the Consiglio Generale were fond of pointing out, "water is one of the four elements which are essential for life."[39]

A dearth of water is generally thought to have hindered the development of the important wool industry, placing Siena at a competitive disadvantage with her more hydraulically favored neighbors. The constitution of 1262 indicates that the government was already concerned with the limitations that the shortage of water placed on the textile industry's potential growth and was willing to undertake the necessary expansion of the water system. A concern with supplying a sufficient quantity of water to industrial users remained an integral component of Sienese water policy. One of the original civic fountains, Fonte della Vetrice, was ceded to the Arte della Lana in 1306; the overflow from Fonte Branda fed industrial *piscine* (pools) and powered the city's mills.[40]

Even a city on the bank of a river was not always guaranteed an adequate supply of potable water. Kingston-upon-Hull, situated on the tidal mouth of the Humber, suffered severe seasonal shortages. According to a series of royal commissions in the later fourteenth and early fifteenth centuries, there was no fresh water running into the town. Water brought in by boat proved a crippling expense to the poor residents, forcing them to leave the town during the summer. Hull was given royal permission to build freshwater dikes (the Julian Dike and the Bushdike), but these caused flooding and resulted in serious disputes and lawsuits with the neighboring landholders. Furthermore, the water in the open ditches was easily polluted, not only from the salt Humber at high tides but also by straw, refuse, and carrion. An underground pipeline seemed the best solution to these difficulties, and in 1447 the town obtained a royal license to acquire springs and convey water to the town by subterranean lead pipes.[41]

Hull's problem with the pollution of its open channels raises the issue of water quality. The reduction in the incidence of waterborne diseases such as cholera, typhoid, infectious hepatitis, bacillary dysentery, and diarrheal disease through improvements in water quality is one of the most compelling modern arguments for replacing traditional water sources with advanced hydraulic systems.[42] Although medieval Europeans lacked a scientific understanding of the role of waterborne pathogens in the transmission of disease, there are indications that contaminated water was thought to cause illness, and considerations of water quality may have played a role in some decisions to adopt complex urban water systems.

The generation of urban waste products increased along with the demographic and industrial expansion of twelfth- and thirteenth-century cities: the more vigorously a city grew, the more acute the problem of waste disposal became. River pollution had four main sources: domestic rubbish either dumped there deliberately or washed in from the streets, animal dung from streets and stables, sewage, and occupational wastes. Well water was liable to be contaminated by the practice of disposing of wastes in rubbish pits and cesspits, which typically honeycombed the yards of medieval tenements. In London, numerous measures were enacted in the attempt to control the most egregious sources of water pollution, but even if they had been strictly observed, the quality of water in the Thames and permeable gravels would have been compromised by practices that were officially condoned, such as drains and latrines that discharged their effluents directly into the river.[43]

Medieval Europeans seem to have been aware that ingesting polluted water could result in illness, even if they lacked a scientific theory of waterborne pathogens. In 1374 Katherine Bishop of Norwich accused Ralph Crete of causing her water supply to become polluted with dung and filth: she claimed that as a consequence of tasting the water and using it in the preparation of food, she and her servants had become so ill that their lives were threatened. Joan Boys of Bamburgh was "poisoned" and prematurely delivered a stillborn child as a result of drinking water from a well that contained the body of a dead dog. Pollution of the *bottino* was believed to be responsible for an outbreak of illness among those drinking water from the Campo fountain in Siena. Odors emanating from polluted water were also thought to cause disease: in 1290 the White Friars of London complained to the king that the putrid exhalations from the Fleet were so powerful that they were rising above the scent of incense burnt upon the altar and had caused the deaths of many of the brethren.[44]

Contaminated or poisoned water was a suspect during plague outbreaks, although in this case the correlation between disease and water pollution was false. The London conduit accounts for 1348–50 include the cost of examining the conduit "when it was slandered for poison," a charge almost certainly associated with the outbreak of Black Death, which arrived in London late in 1348. The dates of London's fourteenth-century sanitation ordinances, which include strictures against polluting the Thames, coincide with the recurring outbreaks of the pestilence in the city. The terrible accusations leveled against Jews or lepers of poisoning wells, particularly prevalent during plague outbreaks, again point to a perceived linkage between water quality and the transmission of disease.[45]

Civic authorities were not indifferent to the problem of water pollution, but fetid odors emanating from filth in the streets were considered the more urgent health hazard. In their ongoing battle to keep the streets clean, municipal governments often tolerated or even prescribed the disposal of refuse in watercourses. Complaints about the accumulation of refuse on the Thames frontage precipitated ordinances calling for the use of dung-boats: these did not eliminate fluvial pollution, but they mitigated it by conveying the filth out to the middle of the river, where the current was strongest.[46] The quality of river water was compromised not because medieval civic officials were insensitive to hygiene but because watercourses were irreplaceable components in sanitation systems that aimed, above all else, at keeping the streets reasonably clean. Though municipal officials continued to issue proclamations against the more egregious sources of river pollution, the problem of providing a clean water supply was tackled more successfully through the adoption of a technological solution. Like the monasteries, medieval cities addressed the conflicting demands of water supply and waste disposal by the creation of artificial intake systems. Rivers and watercourses, fed by surface or subterranean drains, remained key components of the discharge system, but a new source, springwater, was supplied through pipes and distributed in public fountains.

Fetching water from an inconvenient source is costly in terms of time and energy. The aim of improved access to water also seems to have played a role in some adoption decisions. Siena's early fountains were originally situated outside the urban nucleus. The location was dictated by the unusual nature of the main component of the water-collection system, the network of *bottini*. Since the town was initially constructed on the highest parts of the hilltops, the early fountains were left in an exposed position outside the walls, although in the course of the thirteenth and fourteenth centuries, many of the "extramural" fountains became incorporated within the expanded circuit of the city walls. The commune built and paved new roads in order to make the fountains more accessible. Nevertheless, residents of the upper parts of the city had to walk down steep streets to reach the city's thirteenth-century fountains. The climb back up, carrying a heavy water jar or load of laundry, was arduous, and there were reports that women were being assaulted on their way to fetch water. In the mid–fourteenth century, the commune invested thousands of lire in bringing water to a new public fountain in a more central location, the Campo. Its success in bringing water to the Campo's relatively high elevation precipitated a spate of petitions for new fountains to serve *contrade* (neighborhoods) that lacked easy access to the older fountains. Many of the petitioners sought to utilize the Campo overflow for new neighborhood fountains. At least some of these proposed fountains were built, but they seem to have been much smaller than their thirteenth-century predecessors: few seem to have had subsidiary basins, perhaps because the supply of water was inadequate. What the new fountains lacked in water quantity they made up for in increased availability. They do not seem to have been able to supply water for as wide a variety of uses as the older fountains, but they did provide more convenient sources of domestic water for the inhabitants of their neighborhoods.[47]

One argument advanced in Sienese petitions for new fountains was the need for a neighborhood water supply as a protection against urban fires. A 1356 petition from inhabitants of the *contrada* of Abbazia Nuova to build a fountain near the kilns of the potters stressed the need to fight fires; the inhabitants of the neighborhoods of Salicotto and San Salvatore also claimed to need a new fountain not only to supply water for daily use but also for fire protection. The threat was certainly real: some sixty-four fires are known to have broken out in Siena during the fourteenth century alone. Siena's firefighting strategy depended on the city's fountains: when a fire broke out, men and women raced to the nearest fountain and filled earthenware vessels with water. The large fountain basins were designed to allow many individuals to fill their vessels simultaneously. These "water bombs" were apparently hurled into the flames: in the aftermaths of urban conflagrations, the *Biccherna* (city treasury) registers record compensation payments for the hundreds of vessels of various descriptions that were broken in the battle. Besides paying for the broken vessels, the Biccherna also paid the men and women who had acted as emergency water carriers.[48]

The desire for prestige and competition between communities can encourage the adoption of innovations. Civic pride and civic identity could be expressed by the construction of a fountain, and highly decorated fountains, such as Perugia's Fontana Maggiore, stood in core civic spaces. Siena's fountains played a role in her rivalry with other Tuscan cities: "Your city has always been the most beautiful and the most clean of any in Tuscany and possessed of the most beautiful fountains. For this reason, all the foreigners who visit it want to see the Fonte Branda." The Campo fountain (Fonte Gaia) was associated with the Virgin Mary, Siena's patron. The cost of candles dedicated to her at the fountain recurs in the financial accounts of the *operaio dell'acqua*.[49]

English fountains were incorporated into civic displays, serving as decorated stages for symbolic pageants and running with wine during celebrations of important events. On the birth of a son (the future Edward III) to Queen Isabel in 1312, for example, London's "conduit in Chepe ran with nothing but wine for all who chose to drink there." The city welcomed the victorious Henry VI back from Agincourt in a procession that made ceremonial pauses at the conduits en route: Cornhill conduit was decked with red and had a company of chanting prophets who released sparrows and other birds as the king passed; the tower of the Cheapside conduit was green, with a pageant of apostles, kings, martyrs, and confessors of England, who offered the king thin wafers mixed with silver leaves and a cup filled from the conduit pipes.[50]

Once adopted, civic water systems were generally very popular: no sooner did residents become familiar with the new systems than a demand for expanded services arose. By the fourteenth and fifteenth centuries, the distribution pattern shows pockets of intensive secondary diffusion in some towns, such as Bristol and Viterbo, where multiple conduits supplied numerous distribution points. The provision of water for collective systems also stimulated a demand for private branch pipes. Private pipes are known from the thirteenth century on. To begin with they were generally a mark of special privilege. In July of 1244, Henry III granted Edward, son of Odo, the right to take a private pipe the diameter of a goose quill to his court in Westminster. As warden of the king's works at Westminster, Edward had overseen the arrangements for making the conduit at Westminster Palace and had recently been reimbursed for a large sum of money that he seems to have advanced to cover its cost. The branch pipe, which would have derived its water from the new conduit, made a fitting gift from a grateful king. In Paris, once the precedent had been established in the mid–thirteenth century, requests for private branch pipes multiplied, as the local nobility and others with friends in high places used their clout to obtain their own water pipes at the expense of the public supply. Persistence could pay off. Casting covetous eyes on the Lichfield Cathedral conduit, Jordan, the archdeacon of Chester, badgered the dean and the chapter until they granted him permission to attach a small pipe to their main line and run it to his house on Beacon Street in 1280. The reluctant grantors did their best to limit the damage: the archdeacon was allowed to open his pipe only when he was actually at Lichfield, and "when he retires or dies this pipe shall be removed and no successor of his shall have any claim to it." By 1300 the canons' houses in the Close were being supplied by piped water, though as the bishop discovered, some branch pipes were delivering water more abundantly than others. In the early sixteenth century, the chapter decided to grant vicars, chantry priests, and even choristers water from the Cathedral aqueduct.[51]

The Dublin city records show how branch pipes could proliferate. Dublin's civic conduit had been completed in the period between 1244 and 1254. In or about 1254, the city granted water from the civic system to the Priory of the Holy Trinity, the Church of the Holy Savior, and the Friars Preachers, so the precedent of branch pipelines was established almost at once. In 1288 the knight Sir Richard of Exeter had recently obtained the right to take a small private pipe from the city pipe up to his premises, a privilege that his son and heir (also named Richard) then transferred to Henry le Mareschal (mayor in 1281). According to the city confirmation of the transfer, a portion of the supply was to be allocated to the use of the neighbors. (In exchange, Henry was to present the mayor with a chapelet of roses every year on the Feast of Saint John the Baptist.) Two years later, the city made another private pipe grant, this time to William le Deveneys, the town clerk. In 1323 Master Walter de Istelep obtained the right to take a pipe the diameter of a goose quill to his house, in exchange for a rent of six pence. A few years later, Nicholas Falstolf and his wife Cecilia obtained a subsidiary branch pipe from Walter de Istelep's cistern for the rent of one penny, for the use of their own home. The enterprising Falstolfs were also permitted to supply all of their tenements in the parish with their private pipeline.[52]

Illicit extensions of public systems indicate that hydraulic knowledge was filtering down to a broad segment of the population and was stimulating some unregulated "private adoption decisions." In 1478 London's William Campion was hauled up before the mayor and the aldermen for unlawfully tapping a public conduit pipe and conveying the water to his house on Fleet Street and elsewhere. Those responsible for deciding upon his punishment displayed considerable ingenuity and dramatic flair. The malefactor was set on a horse and led through the streets of the city with "a vessel like unto a conduit full of water upon his head, the same water running by small pipes out of the same vessel," while his crime was publicly proclaimed. When the water had run out of this conduit-hat onto the (probably) humiliated and (undoubtedly) damp Campion, the vessel was refilled.[53]

Even some villagers were displaying considerable technical enterprise in the matter of illicit water diversions. Sir Reynold Hagbech brought John Snake and his fellow villagers at Emneth (Norfolk) to the king's court on a charge of trespass in 1387. The villagers had diverted Sir Reynold's stream and were distributing the water in turn among their dwellings through a subterranean network of lead and wood pipes—a practice that the villagers freely admitted but which they defended on the grounds that everyone in the village had done so from time immemorial.[54]

Waste disposal also stimulated some creative private innovations. In 1314 Alice Wade was summoned before the mayor of London for running a wooden pipe from her indoor privy and connecting it to a subterranean public gutter, which passed beneath the street and the houses in Queenhithe. Wade's clever attempt to hook up her latrine to a public drain was discovered when the filth she had been casting down her privy clogged the system so that it began backing up and fouling the houses in the neighborhood. She was given forty days to remove the pipe.[55] We know about the hydraulic high jinks of William Campion, John Snake, and Alice Wade only because they got caught: how many other secret branch lines remained hidden from the eyes of the authorities? The public spectacle of Campion's conduit-hat may have deterred some potential pipe-tappers, but it also would have helped advertise the idea of secretly filching water from the public mains.

Some towns were able to minimize the disadvantages inherent in adopting civic water systems by directly collaborating with their urban religious houses. By joining forces with the more hydraulically experienced religious sponsors, cities were able to acquire access to water at a reduced cost and risk. The Paris municipal supply depended on the monastic systems of Saint-Laurent and Saint-Martin-des-Champs. The Dublin civic water supply of 1244 seems to have originally been some sort of joint venture with the abbey of Saint Thomas the Martyr, and the city subsequently granted piped water from its supply to four other religious houses. In 1310 the Franciscans in Southampton shared the water from their conduit with the townsmen: a pipe fed a stone basin situated outside the walls of the friary. By the terms of a 1346 deed, the water from the dean and chapter's conduit in Exeter was split into three branches: one supplied the cathedral, one the priory of Saint Nicholas, and one the city. Most of these arrangements seem to have worked fairly well, but the Franciscans at Newcastle-upon-Tyne came to regret their decision to share their conduit with the townsmen. According to a complaint made to the king in 1341, the friars had permitted the city to share their abundant conduit-head, which was enclosed in a stone conduit house with a locked door. The quarrel seems to have stemmed from the friars' decision to keep the key for themselves. The townsmen broke down the conduit house door by force, and (at least according to the friars' side of the story) broke the friary conduit and diverted its water unjustly.[56]

In the late fourteenth and fifteenth centuries, more complex agreements were reached between religious institutions and townsmen for joint sponsorship of water systems. According to an agreement made in Bristol in 1391, the town granted the Dominicans a "feather" from the civic Key-pipe in exchange for the friars' conduit, spring, and lead pipes. The friars were guaranteed a sufficient supply from the civic system and specifically exempted from maintenance costs; the city, for the cost of a branch pipe and an annual payment of a mere twelve pence, obtained a ready-made addition to the town supply.[57]

By the end of the century, the Southampton friars were having problems maintaining their system, and in 1420 the warden and the friars transferred responsibility for the system to the borough. According to the terms of the agreement, the city was to take up and recast the old lead pipes and lay them along the original line. A new stone water-house was to be built next to the friary: here the water was to rise in one large lead pipe and be divided into two identical branch pipes. One of them would serve the friary, and the other would feed a cistern that was to be for the benefit of the town. The friars retained a key to the new water-house and control of the conduit-head and were to admit representatives from the city as necessary. If, through the friars' default, the system was damaged, the city officials could break into the stone water-house and take the whole main pipe into their possession. Both parties benefited: the friars got their decaying system repaired and future maintenance subsidized, and the town was able to utilize an existing conduit-head and its spring and obtain access to an existing route. In an analogous agreement of 1438, the burgesses of Gloucester acquired the rights to three-quarters of the Franciscans' piped water supply and the right to extend the system as far as the High Cross. As in Southampton, the agreement harnessed the operational expertise of the friars and the financial resources of the townsmen in a new joint venture beneficial to all parties.[58]

The presence of other local hydraulic structures might also encourage adoption decisions. In Italy archaeological surveys have shown that segments of old Roman aqueducts were refurbished and incorporated into new medieval systems at Narni, Spoleto, and Cività Castellana. Several private fountains, purchased by the city, formed the original core of Siena's public water system.[59] Cities lacking such a local nucleus for a conduit rarely took the riskier and more costly step of building an entirely independent civic system. The technological style and trajectory of the derivative systems were largely determined by the preexisting components they incorporated. English municipal systems utilized the conduit houses and subterranean pipes characteristic of local monastic systems. Siena's public water system, in contrast, developed a system of fountains and filtration conduits that seems to have been based on the technological attributes of the local fountains it acquired.

The available evidence suggests that the adoption of complex water systems by medieval communities followed a common pattern: dissatisfaction with existing conditions, coupled with a recognition of specific problems and needs, created a social climate potentially receptive to technological innovation. Monastic houses and civic authorities could be persuaded to adopt complex water systems if they felt that the perceived advantages significantly outweighed the perceived disadvantages. Judging from the problems that arose when complex water systems were adopted, as will be seen in the following chapters, the disadvantages are likely to have included high costs; the necessity for land acquisition; damage to streets, structures, and property incurred during construction and repairs; and administrative and maintenance headaches.

The new hydraulic systems helped mitigate the impact of demographic and industrial growth on fluvial and groundwater pollution by providing an alternative, cleaner supply for domestic consumption. Urban fountains were a convenient source of good-quality water; they were desirable and prestigious urban amenities. Monastic conduits helped enclosed communities meet the ritual and practical requirements of a closely regulated way of life. Nevertheless, complex water systems were expensive to build and needed constant maintenance. On the whole, the perceived benefits and perceived disadvantages seem to have remained more or less evenly balanced: inertia tended to inhibit the spread of hydraulic technology, but occasionally a strong recognition of new needs or the opportunity to incorporate an existing system could tilt the scales in favor of adoption. Although a significant minority of medieval communities did decide to take the risk of building complex hydraulic systems, such a decision was by no means a foregone conclusion: many medieval monasteries and most European cities remained dependent on traditional water sources well beyond the Middle Ages.

2 Resource Acquisition

The potential sponsor of a medieval conduit had to solve two immediate problems: the acquisition of a supply of fresh water and access to a continuous strip of land with a suitable gradient between the source and the destination. The configuration of a system had to be adapted to both the physical topography and the social landscape. Any potential conduit route was likely to cross lands held by at least one landowner, and quite possibly several—the conduit of Waltham Abbey crossed so many properties that thirteen separate charters pertaining to its route were drawn up and preserved in the abbey's cartularies.[1] The most direct expedient was the outright acquisition of the necessary land and water sources, through donation, purchase, or appropriation. A second solution was to obtain an easement, either to utilize another's water source or to carry a conduit through someone else's land. Like other medieval land transactions, these agreements were often formally recorded and preserved to guard against challenges or disputes. Charters recording the agreements between landholders and conduit sponsors, together with licenses to construct conduits, are among the most abundant classes of documents pertaining to medieval water systems. Often they are the only surviving textual records.

If preliminary negotiations with local landholders were encouraging and the potential sponsors decided to go ahead with their hydraulic project, formal land agreements would have to be obtained. Once the general line of a route had been selected, detailed arrangements would have to be made with each landholder, both for the initial construction of the conduit and for subsequent access for inspection and repairs. Such arrangements might not have posed a serious problem for a monastic house in the "wilderness," if the desired route crossed the property of a single large landholder, but for an urban sponsor they could be a matter of some complexity. Even a short length of pipe could easily cross multiple tenements. The fourteenth-century pipeline to the cathedral close at Lichfield, for example, had to pass through the properties of William de Harperly, Robert Cooke, Robert de Knyttcrote, and Agnes Sparham, each of whom held lands near the west gate of the close. The Friars Minor of Cambridge had to purchase narrow strips of land from seventeen persons for the construction of their aqueduct in 1325.[2]

The actual construction of the new system would normally have commenced sometime after the land transactions had been completed. The charters and licenses themselves can often be dated with precision, but they provide a definite date only for the land acquisition stage of what could be a considerably longer adoption process. In some cases the construction of the conduit through the designated land seems to have quickly followed the drafting of the charter. At Waltham Abbey, at least some of the charters were apparently drawn up after work on the initial sections of the conduit had already begun. These last-minute charters all refer to lands that were crossed in the later stages of the construction campaign. Either the canons at Waltham were unusually complacent in trusting that their informal preliminary agreements with the landholders would be honored, or the conduit engineer, Laurence of Stratford, decided upon a change in the route once work had started.[3]

In other instances, some years elapsed between the date of the grant and the construction or completion of the conduit. A series of charters granting water and land for an aqueduct for Cirencester Abbey seem to date to the early thirteenth century, but two others were apparently drawn up in the latter part of the century, which suggests that the project fell into abeyance for several decades. Gilbert de Sanford granted land and springs to the city of London in 1237 for the construction of the Great Conduit, but, according to the London annals, work on the conduit did not begin until 1245; the date of its completion remains unknown. In Southampton, Nicholas de Barbflete granted a spring for the friary conduit in 1290, but the friars did not secure a license to lay their pipes until 1327. Without supporting evidence, a land transaction alone cannot provide a secure date for a conduit, nor does it prove that an intended conduit was ever actually completed.[4]

Often the land involved in these transactions was restricted to the amount actually required for the components of the water system. In many cases, the grants allowed the builder flexibility in the final choice of land within the grantor's larger holding. Alice Chobham's grant to the city of London allowed the city to have a plot of land twenty-four feet square for a spring "wherever they may choose" within all her land "atte Cherchende." Gilbert de Sanford's grant to the citizens of London permitted the city to bring the conduit through "such parts of his fief as they deemed expedient."[5] Such grants made practical sense: there was no point in expending resources on specific route selection before access to the general line of a proposed conduit had been secured throughout its entire length. Once all necessary permissions had been obtained, the final details of ascertaining levels and choosing the route could be left to the engineer, who, thanks to grants such as these, seems to have been allowed a fair degree of latitude in his task.

The impetus for land transactions almost certainly came, in most cases, from the conduit's sponsors. Occasionally the donation charters preserve a record of the original request. Robert de Berkeley's grant to the church of Saint Mary Redcliffe at Bristol took place at the request of William, the church's chaplain. In London, the head of the Greyfriars' aqueduct was granted to the friars by William, tailor to King Henry III, following a request by William de Baysynges, the brother in charge of the works.[6]

Between the complexities of land tenure and the desire to forestall future challenges, the prudent patron of a complex water system would not only have to make sure that satisfactory agreements were reached for each property en route but also might have to obtain more than one document for a single piece of land. As a result, those segments of society most at ease with the intricacies of medieval red tape would be least intimidated by the process of land acquisition. One of the problems was that the grantors themselves were not necessarily absolute owners of the land in question; they might be tenants in a chain of subinfeudations.

Waltham Abbey's conduit charters illustrate the kind of tenurial tangles that a conduit sponsor might have to unravel. Richard Picerne, with the consent of all his free men of Wormley, granted the Waltham canons a license to take their conduit across forty-two perches of marsh that he held jointly in demesne with Alexander of Pointon. He also granted Waltham forty perches of the common marsh pertaining to their men there, twenty-four perches of the land of his man Richard, and forty perches of the land of his free man Henry, son of William. The crossing of the jointly held marsh was confirmed in a separate charter by Alexander of Pointon, also acting with the consent of his free men at Wormley. Alexander also permitted the canons to take their conduit over the land of John and across forty-four perches of his common pasture. Henry, son of William, was himself the grantor in three other charters; since he was Richard Picerne's free man, the grants were made with Richard's approval. A grant by John of Stewkley was made with the consent of his lord Robert, son of Humphrey. Henry of Crossbrook and William the miller made a grant on their own behalf (witnessed by Richard Picerne), as did Thomas of Haverhill, Walter de la Hale, Richard Hook, and Nicholas the clerk. The final grantor, Henry son of William Portingale, was under the lordship of the canons.[7]

Since landholding was largely hereditary, the grants generally stipulate that the heirs and assigns of the grantor will be bound by the same agreement in perpetuity, a common enough clause in medieval land transactions but nonetheless an important protection for the builder of an expensive conduit. To guard against the possibility of future legal challenges, confirmations by other family members might be sought when the original charter was drawn up. For example, the confirmation by John, son and heir of Arnold of Bagendon, of his father's grant to the canons of Cirencester of a piece of land next to the spring called Letherwell bears the same witness list as his father's original donation, which suggests that the two documents were drafted on the same occasion.[8]

The growth of centralized administration in England led, by the thirteenth century, to a fairly standardized sequence of steps when permission to take a conduit across royal land was sought. In these cases the king might order an initial inquisition *ad quod damnum* to determine the extent of potential damages to royal interests. A writ would be sent to the sheriff or other appropriate local official to order the inquest, which was conducted by a body of jurors. If the results of the inquisition were acceptable, a royal license would be issued and enrolled by the Chancery and a record preserved in the Patent Rolls. Under Edward I, for example, licenses for making conduits were granted to the convent of Saint Werbergh, Chester, and the Friars Minor of Northampton after inquisitions *ad quod damnum* made by the justice of Chester and the sheriff of Northampton, respectively. These inquests were particularly, but not exclusively, concerned with damages to the royal demesne: the licenses issued following the inquisitions contain provisions for repair of damages to lands of "others" as well as lands of the king, along with provisions pertaining to other royal concerns, such as highways and city walls.[9]

In Dublin a writ for a similar inquisition was issued by Maurice FitzGerald, justiciary of Ireland in 1244. He commanded the sheriff of Dublin, by twelve free and lawful men of his county, to make an inquisition, with the advice of the mayor and the citizens, as to where water could best be taken and conducted to the city. The same twelve men were to ascertain whether any damage could arise in the process, and the results of such inquiry were to be returned under seal to the justiciary. In this last case the jury was charged not only with the normal inquisition concerning potential damages; it also apparently exercised some discretion in the selection of the water source and conduit route.[10]

Royal confirmation of a private grant was not usually necessary, though in certain circumstances it might be desirable. In Bristol the Friars Minor petitioned Edward III for a confirmation of a spring and conduit-head granted to them in the time of his grandfather, Edward I. The exact circumstances behind this request are unclear—perhaps doubts had arisen concerning the validity of the friars' title to the water supply. In any case, the confirmation they sought was granted to them in 1374. It was also possible to obtain subsequent royal confirmations for royal conduit licenses, as demonstrated by Edward II's and Henry VI's confirmations of a license to build a conduit through the king's meadows in Hinksey, originally granted to Oxford's Blackfriars by Edward I.[11]

Religious houses often procured desired lands by means of pious donations. When Richard Oseney and his wife Agnes granted the Cathedral priory of Worcester the right to take a conduit through their meadow, they asked in exchange that following their deaths, the prior and convent would accept their bodies for burial in the cathedral, ring the bells, say a requiem mass, and pray for them as special benefactors. Some donations may have more complex background circumstances, although these are not usually apparent in the donation charters. Philip and Isabel Burnell of Malpas gave a spring, a plot of land around it, and the easement for a pipe to the Abbey of Saint Werburgh in Chester in 1282, in what appears to be a simple pious donation. The Burnells, however, had won a lawsuit against the abbot the previous year, and the abbot had been forced to pay £200 to retain several manors that Isabel claimed as her inheritance. Evidently the Burnells had come to the conclusion that their salvation was more important than the money, because they remitted the £200 on the condition that the monastery provide two chaplains to say perpetual prayers for Philip. It would appear that the gift of the spring and the easement in the following year was part of their attempt to mend fences with the abbey in the aftermath of their lawsuit.[12]

The spatial distribution of pious generosity may not always have followed the desired topographic contours. Four of the five surviving Farfa Abbey land transactions for the construction of its eighth-century aqueduct are, apparently, straightforward donations, but in the fifth the abbey resorts to an outright exchange of land. Waltham Abbey received from Henry, son of William of Wormley, two separate grants of land with springs for its conduit. The first seems to be a simple donation "for the salvation of myself, my ancestors, and my heirs." The second, however, although still "for the salvation of myself, my wife, my ancestors and my heirs," contains the explicit provision that "in consideration of my homage and service and in recompense for my gift," the church and canons convey to Henry and his heirs "that land called Priest's Acre" and a meadow. Both Farfa and Waltham seem to have pursued mixed strategies: they were happy to accept donations when they could get them but were prepared to bargain when the promise of spiritual reward was not enough.[13]

Urban friaries were beneficiaries not only of traditional donors, the rural landholders, but of a new, urban class of patrons, merchants and artisans who, in spite of their urban base, might also hold land near the town. One such was Henry the bell-founder, a donor who, with a townsman's shrewdness, sought to ensure that his generosity was not frittered away. His grant in perpetual alms to the Friars Minor of Lichfield of springs and permission to construct a conduit house and lay pipes in his land was made on the strict condition that the friars give no vessel of the water to anyone else without first obtaining his special permission.[14] Small grants of land, and particularly of easements for conduits to religious houses, were popular with all classes of landholders, one suspects, because they constituted an inexpensive form of piety. Given sufficient guarantees against damages to his land, the donor's bequest, though holding out the promise of spiritual reward, actually cost him very little in terms of his earthly goods.

A side effect of the thirteenth-century mortmain restrictions in England was the creation of a new hurdle for the sponsors of ecclesiastical water systems. Walter, son of Thomas Toky, needed a license for alienation in mortmain to grant water and land for an aqueduct to the Carmelite friars of Gloucester. The Friars Minor of Lynn found their water supply threatened by their failure to observe the mortmain statute. They had acquired a water source from Thomas Bardolf and Robert de Scales without first obtaining a license from Edward I, in contravention of the statute. In spite of this omission, Edward II in 1314 granted them a license to retain the water source and build a subterranean conduit to their house, but only after the sheriff of Norfolk had issued a favorable report following an inquisition *ad quod damnum*. The tendency to grant easements for the use of springs and land for pipes rather than directly alienate the land itself may have accelerated as a result of mortmain legislation.[15]

In some cases, grants of water sources were the result of direct financial settlements. Siena pursued a policy of actively soliciting voluntary sales to the commune of private properties that contained fountains or were adjacent to them: a series of such purchases, made on behalf of the city by Ildibrandino Bolgarini in 1221, included the fountain of Val di Montone and the zone around Fonte di Follonica. The conduit to Lichfield Close was fed by several springs at Pipe, two of which were sold to the church in the mid–twelfth century by Thomas Bromley for 15s. 4d.; another was given by William Bell in return for 12s. a century later. In 1355 Alice, the widow of William Chobham, granted to the mayor and the commonalty of the city a plot of land for one spring for the London conduit, in return for 3d. sterling, to be paid annually at the feast of Saint Michael to herself, her heirs, and her assigns in perpetuity. Permission to take a pipe across another's land could occasion a regular payment. At Canterbury, William son of Drogo received one penny each year from the monks of Christ Church "for our aqueduct which passes through his land." John Hull, in 1420, granted the city of Exeter permission to take a pipe through his land upon a payment of 8d., to be paid at the feast of Saint Michael, for the easement. Even the king paid a compensation of £1 6s. 8d. to Argentine, the widow of Master Alexander the Carpenter, since the pipe for the lead bath for the royal falcons in the Mews at Charing Cross passed through her land. Such a substantial sum probably indicates a comprehensive settlement rather than an annual payment.[16]

The cases dealt with so far were essentially voluntary agreements. On other occasions, however, some degree of compulsion may have been involved. Some royal licenses for conduits permitted the conduit builder to take his water supply through not only the king's lands but the lands of unspecified "other persons." The Friars Minor of Northampton were permitted to take a subterranean conduit to their house provided that they indemnify the persons who held land in the field en route: it is not clear whether or not these persons had previously agreed to let the conduit pass through their land. The commune of Siena, fortified by the legal concept of public utility, seems, on occasion, to have expropriated land needed for water supplies and other public works by the expedient of compulsory purchase. In the constitution of 1262, provisions were made for building a new fountain. The land with the desired vein of water was to be purchased from whoever held it. The price would be determined by two appraisers, one chosen by the landowner and the other by the men of the neighborhood.[17]

The royal authority of the English king could override the objections of landholders. In the case of the Dublin conduit, the jurors were to select the best route: any damages would be repaired at the cost of the king, but opposition from the landholders was evidently considered a distinct possibility in the eyes of the justiciary. In his writ to the Sheriff of Dublin, he commanded that any who opposed were to be suppressed by force and attached to appear at the next Assizes; those who resisted were to be arrested and held till further mandate. In Chester, forester Randle de Merton cut the newly laid pipes of Saint Werburgh's Abbey where they passed through his land. Since Saint Werburgh's had obtained a royal conduit license two years earlier, which permitted them to take their pipes from the spring through any intervening estates to their house, Edward I ordered Randle to repair the pipes and make compensation.[18]

Siena's network of *bottini* (subterranean filtration conduits), which supplied the city fountains, was constantly being expanded. Since the workmen followed promising veins of water as they dug the tunnels, it was inevitable that the bottini would pass under private lands and that the ventilation shafts (*smiragli*) would, on occasion, have to be situated in private property. In order to expedite the expansion of the bottini network, the civic authorities used their political and legal power to compel landowners to allow bottini to be taken through their properties, although provision was made for compensation in case of damages. Nevertheless, the obstruction of works by hostile landowners seems to have been enough of a problem to require strong legal sanctions: impeding workmen engaged in aqueduct construction was punished with the heavy fine of 100 lire. The disruption caused by the movement of men and materials during the construction process was not the only source of dissatisfaction. According to a sixteenth-century statute, once the bottino was built, the zone around any openings had to be left permanently free of trees and remain uncultivated for a distance of four *brachia* in all directions, presumably to protect the water supply from pollution and to guard against root damage to the conduit.[19]

Occasionally, opposition to conduit construction could turn violent. In spite of a royal license issued in 1380, the Sudbury Dominicans encountered serious local opposition in the course of building their conduit. In 1385 the king issued a public proclamation of protection for the friars' men, servants, and laborers, who found themselves in peril "at the hands of certain rivals" who were hindering the works. Kingston-upon-Hull's attempts to supply the town by means of a freshwater dike in the late fourteenth and early fifteenth centuries were constantly obstructed by the inhabitants of the villages nearby. The opposition was so persistent that some of the objectors were eventually hanged in York. Even the Roman curia was dragged into the Hull dispute and issued a call for cooperation in 1412.[20]

Whatever methods were used for acquiring access to land for the construction of a conduit, provisions for continuous access were necessary for inspections and repairs. A common feature in many of the English grants is the provision for free ingress and egress to the pipeline for these purposes. Gilbert de Sanford's 1237 grant of springs in his fief at Tyburn for the London conduit included the provision that neither he nor his heirs could at any time hinder the citizens if they wished to open or dig up the conduit-head or the pipes. A fifteenth-century grant of springs at Paddington to the city by the abbot and convent of Westminster permitted access to the water system, but only by existing roads and paths.[21]

Subsequent changes in land tenure might threaten these safeguards, however. The twelfth-century pipe of the Christ Church, Canterbury, aqueduct passed through a garden that, at the time the pipe was laid, belonged to the archdeacon. Between February and May of 1227, this land (with the explicit exception of the watercourse) was granted by Archbishop Langton to the priory of Saint Gregory. This alienation of the land through which their pipe passed, even if made to a friendly house, apparently caused the monks of Christ Church some anxiety. Although the prior and convent issued an *inspeximus* (official copy) of the grant, they seem to have sought additional protection for their rights to maintain their water system. In July the prior and convent of Saint Gregory's reassured the prior and convent of Christ Church that the conduit would remain unharmed and that workmen would be permitted access to it in their orchard whenever necessary. In Dublin, the pipe of the Friars Preachers passed through lands that had passed from the possession of William the Clerk to that of the mayor, Roger de Asshebourne. A new grant was issued by the mayor, authorizing the friars to dig in his lands and mend their pipe as often as necessary, but they were not to remove it from the ground that they had in the time of William the Clerk. This last clause was probably designed to protect Roger's other holdings against any new pipe trenches by restricting the easement to its original location.[22]

Such guarantees of access were important, for the excavation of trenches during both construction and repair works could damage crops or cause other losses to the landholder. Farfa prudently obtained explicit permission from each donor to remove any trees that happened to grow along the course of the aqueduct. Two water-supply grants by William Geraud of Gloucester, one to the Abbey of Saint Peter and one to the Friars Minor, reveal the issues that were a source of anxiety to this particular individual: damage arising from overflowing of the watercourse and damage to his grass and meadow because of the conduit. The inquisition *ad quod damnum* for the friars' conduit at Northampton estimated that damages of one mark would be sustained by the holders of a field "if the land was sown at the time."[23]

If damage to the landholder's property did occur, procedures for compensation were sometimes spelled out. Indemnity clauses basing damage compensations on the assessment of a specified number of local lawworthy men are a common feature of English conduit charters. A similar concern for damage compensation can be seen in Italy. In Viterbo, the city allowed the residents of the *contrada* (ward) of San Faustino to build a fountain and an aqueduct, as long as they made amends for damage caused by the digging, according to the judgment of two legal men. Siena's communal government kept its promise to pay compensation for damage: the *Biccherna* (city treasury) accounts contain many examples of payments to individuals who suffered losses resulting from the construction of *bottini* and, in particular, their *smiragli.*[24]

In spite of all attempts to forestall problems, disputes between patrons and landholders still broke out. Lichfield's water supply from the conduit at Pipe was the subject of several quarrels. In 1293 Thomas of Abnall, through whose land the pipe ran, was forced to acknowledge the dean and chapter's right to an uninterrupted supply of water and access to his land to maintain the pipe. The system was again threatened with forced "discontinuance" by aggressively hostile landholders in the fifteenth century: Sir Humphrey Stanley of Pipe cut off the supply effectively (if crudely) by smashing the aqueduct in 1480. When the dean and chapter repaired it, Stanley's wife demolished it again, breaking down the door of the conduit house and damaging the cistern for good measure. The supply was not restored until 1489, when Henry VII ordered the Stanleys to stop interfering with it. In 1439 William Darsett was taken to court for breaking and taking away the lead pipes that passed through a tenement in Coventry, in spite of an earlier grant allowing the city to take the civic conduit pipes under the land. Darsett's action cut off the supply of water, to the great nuisance of his fellow citizens.[25]

It is not always clear whether deliberate acts damaging water systems were instigated by disgruntled landowners or whether they were simply the work of vandals and troublemakers. In 1294 the prior of Daventry accused fourteen men of breaking the priory's underground conduit, carrying away his goods, and assaulting one of his monks and two of his servants. Similarly, someone broke up the underground conduit belonging to the Friars Preachers of Kings Lynn in 1308. Perugia's partially completed aqueduct was vandalized to such an extent that the council called for a commission of ten citizens and the captain of the Popolo to examine the damage and punish the culprits.[26]

In urban areas, pipes were frequently laid under the streets to avoid buildings, though once outside the city walls they could follow a more direct route. The precise route of the London Greyfriars' conduit is known from a copy of a fourteenth-century topographical description of the course of the pipe, preserved in the convent's register. Starting at the convent, the pipe passed under the new wall of the friary and ran along the north side of the street toward Newgate, crossed under the city gate, and ran under the cemetery wall of Saint Sepulchre's churchyard. It then followed the curve of the street, crossed an intersection, turned toward Holborn Bridge, passed under the water of the Fleet, ran westward, along Holborn, to Leather Lane, where it turned north, keeping to the west side of the lane. At the end of Leather Lane it reached open fields, and thence the course ran west again, this time directly across the fields and hedges to a mill and finally to two conduit-heads, the nearer (White Conduit) hidden underground, the farther (Chimney Conduit) visible as "the little stone house seen from a distance" across the open countryside. While in the built-up area of city and suburb, the pipe avoided buildings by following the twists and turns of the streets (although it did pass beneath the occasional precinct wall); once in the open fields, however, a more direct route was possible.[27]

In Canterbury, the Christ Church rentals indicate that, rather than deviating to follow the line of Old Ruttington Lane, the conduit passed directly across several suburban tenements before reaching the priory precinct. In any case, since Christ Church already owned these tenements, it presumably had little difficulty obtaining access to a direct route.[28]

Urban expansion over previously accessible sections of a pipeline could create unforseen problems. The preamble of a royal charter of 1443 concerning the London springs and conduits precisely identifies the difficulties arising from the growth of the city. The king's land at the Mews and others' lands, over and under which the various components of the water supply system were situated, "are lately enclosed by walls and other edifices, so that the mayor, aldermen and citizens cannot examine, clean or repair them without much trouble and difficulty."[29]

Even when a pipe was located under the streets, problems were not entirely avoided: pavements could be damaged and thoroughfares obstructed when it was laid or dug up for repairs. Worcester Priory was allowed to lay pipes under the city streets and the king's highway, but it also had to promise to give reasonable warning when repairs were necessary and maintain the highway over the waterworks. When Dublin's Nicholas Fastolf and his wife Cecilia were allowed to take their private branch pipe through the middle of Rochelistrete to their tenements, they were permitted to open the street for construction and repairs, but only under the condition that such work would be speedily done and that they would repair and repave the street at their own expense once they had finished.[30]

Other parts of the urban fabric could also be at risk. Dublin's municipal authorities granted permission to the Friars Preachers to lay a pipe through the land of the city and across the river, as long as they did not damage the bridge. The Friars Minor at Colchester obtained a royal license to take a conduit into the town by boring through the town wall, provided that they repair the wall at their own cost. Likewise the Friars Preachers at Chester were granted a license to bring a conduit from outside the city to their house, as long as they filled in the king's land, the city wall, and the highways where necessary. In Exeter, the Friars Minor's house lay outside the walls, but the spring for their conduit was inside. So, in spite of their extramural site, they too were faced with the problem of taking a pipe trench under city streets and through the walls. A license, granted them at the bequest of Queen Isabella, allowed them to take their pipe down the middle of Bolehulstrete and either under the south gate or wall or through the middle of the wall of the city, provided that they repair any ensuing damage to the street, the wall, and the gate.[31]

By the late fourteenth century, the city officials of Exeter had a clear grasp of the problems associated with urban pipe trenches and were making elaborate provisions to mitigate them. In an agreement of 1387, the mayor and the citizens granted permission to the monks of Saint Nicholas Priory to dig up the streets and pavements in order to lay pipes for their water supply but stipulated that such excavations must be supervised by four men appointed by the mayor and the community. The monks could dig up the streets when undertaking pipe repairs but had to make them good again within three days of completion of the work. While the pipe trenches remained open, they were to be guarded and covered, so that passers-by would not be injured. If, in spite of these precautions, such injuries did occur, the priory was responsible for providing compensation and would be liable for a penalty of twenty shillings if it was not paid according to the judgment of the mayor and four stewards of the city within fifteen days.[32]

When repairs were necessary, it could be vital to have proof of such agreements. In Cambridge an inquisition of 1434 charged that the Greyfriars, in the course of repairing their pipe, were digging up public land and making openings in streets and lanes in contempt of the king's authority. Moreover, they were called on the carpet for having neglected to obtain a royal license for their original purchases of strips of land for their pipe in 1325, in contravention of the mortmain statute. Although the friars managed to keep their conduit, their failure to produce the proper paperwork left them vulnerable. They had refused a petition from the master and scholars of King's Hall for a branch pipe but were forced to concede a "qwil" of water to them when the King's Hall men armed themselves with Letters Patent from the king in 1441.[33]

In Bristol, workmen from Saint John's Hospital were allowed to mend a broken pipe in the yard of Saint Mary Redcliffe in the face of opposition from the chapel authorities, because of the provisions of the early-thirteenth-century grant. Even with the various grants they had obtained, the Chapter at Lichfield occasionally had difficulty enforcing their right to repair their conduit. In London four servants of the Greyfriars were committed to Newgate Prison for breaking the pavement of the king's highway outside of Newgate. In a plea before the mayor and the aldermen at the Guildhall, the Friars Minor asked for the men's release on the grounds that they had taken up the pavement to mend the channel of the friary conduit (which, as the register description confirms, passed along the road outside Newgate). The friars claimed that these repairs were in accordance with a royal charter permitting them to inspect their watercourse whenever necessary. The court was willing to release the servants on mainprise but ordered the friars to submit their charter to the court.[34]

As conduits became more common, a new problem arose: increased competition for suitable water sources. In the middle of the fourteenth century, a virulent dispute broke out between the Abbey of Saint Peter and the Friars Minor in Gloucester, each of whom had, in the previous century, been granted a supply of water by William Geraud from his lands at Matlesknoll (Robinswood Hill). The friars claimed all the water from the spring called Breresclyft, a claim denied by the monks of the abbey, who in turn charged that the friars were interfering with their own supply. The dispute was only settled when an inquiry was held by Edward the Black Prince, which resulted in a royal license of 1355, specifying that the friars were entitled to one-third of the water from the hill. In 1357 an agreement to this effect was finally drawn up between the monks and the friars. A new reservoir was to be constructed beneath the disputed spring, from which the friars were to have one, and the monks two, lead pipes of equal diameters.[35]

To avoid such potential conflicts, some later medieval grants contained the stipulation that a new conduit was not to occasion loss to an existing water system. In 1390 the mayor and the aldermen of London permitted the men of Farringdon Ward to build a conduit, but only on the condition that it was not injurious to the main city supply, the Great Conduit in Chepe. Six men of the neighborhood were required to give security that if the new conduit should prove harmful to the Great Conduit supply at any future time, the new pipes should be removed and cease to convey water, and the entirety of the Great Conduit pipe should be restored to the condition it was in on the day the license was granted. A grant of springs in Paddington to the city of London by the abbot and convent of Westminster in 1439 provided for reseizer, should the abbey's ancient supply of water from the manor of Hyde be interfered with in any way.[36]

In cases where a conduit seems to follow a suboptimal route from a topographic standpoint, the siting may reflect restrictions on land access rather than engineering incompetence. The municipal watercourse at Dublin followed the shortest route available without encroaching on the territory of Saint Thomas's Abbey. A line through the abbey's lands would have been more direct, but the abbey would not give its permission—hence the watercourse ran a few yards outside the abbey boundary. The fifteenth-century London conduit that was fed by springs in Paddington donated by the Abbey of Westminster was, by the terms of the grant, specifically excluded from running through the abbey's manor of Hyde. This prohibition prevented the utilization of the topographically optimal route (along the present Hyde Park and Piccadilly) and forced the conduit's engineer to take the pipe across intervening higher contours in the area around Marble Arch by means of a "long drain," a trench that must have been more than ten feet below the surface of the ground to maintain the pipe gradient. This more difficult and more costly undertaking may explain the apparent delays in finishing the project (begun in 1431 but not finished until 1471); a less determined sponsor might have abandoned it altogether.[37]

Sometimes the obstacles could prove to be just too much, as Scarborough's Dominicans discovered in 1283. Hoping to run water to their workshops, they sought permission to pull down an obstructing wall, only to be opposed by the burgesses of the town. Moreover, when they petitioned the king to be allowed to draw water from a spring at Gildhuscliff, they were rebuffed. Once again, the citizens of Scarborough had thwarted the Dominicans' hydraulic hopes. Having failed in their attempt to build a conduit on their own, the townsmen had recently granted the spring to the dean of York so that he could build a conduit for the Friars Minor, on the condition that the town should enjoy joint use of the water.[38]

The problems posed by the technological requirement for a physically continuous channel or pipeline, following a downhill gradient from water supply to destination, were solved in the Middle Ages by recourse to various types of land transactions. Direct acquisition provided a simple resolution of the problem of future access for repairs and protected the conduit line for subsequent generations. Easements, however, had certain advantages: they were not as costly (either to a donor or a purchaser), could permit greater flexibility in the final choice of a route, were not subject to mortmain restrictions, and were more acceptable transactions for mendicant orders with strictures concerning the ownership of property. Against these advantages, there was the danger that disputes might arise should damage occur either to the conduit from the landholder or to the land from the conduit owner, as well as the possibility that, should the land be alienated, the new holder might not respect the original agreement.

As these problems were recognized, increasingly elaborate provisions for repairs and damage compensation were included in water grants. Not all disputes were averted, but most arrangements were probably satisfactory. It would not have been in the interest of a conduit owner to antagonize the holder of lands that contained his pipe. William Briewere, Jr., for instance, was evidently satisfied with the way Tor Abbey had treated his father's lands when building their conduit, since he, in turn, donated another spring and watercourse to the same house.[39] Though it is probable that records of some conflicts have been lost, had disputes between conduit owners and landholders been the norm, it is unlikely that the generous provision of springs and access to land would have continued at such a high rate throughout the thirteenth and fourteenth centuries.

The need for land access for long-distance conduits was one of the reasons that adoption of complex systems was restricted to certain social groups. Those in control of land themselves, or those with enough wealth or political power to obtain access to the water and land of their neighbors, were in a favored position to adopt the new technology. In the case of religious houses, many of the older monasteries were wealthy and powerful in their own right. Others, like the mendicant friars, were paradoxically able to obtain access to land by their very renunciation of wealth and power. The irony of begging friars boasting an elaborate and expensive water system was not lost upon the anonymous author of "Pierce the Ploughman's Crede," who scathingly described London's Blackfriars' cloister with its "conduits of clean tin, closed all about, washing basins wrought of shining latten. . . . And yet these builders will beg a bag full of wheat of a poor man."[40] Nevertheless, spiritual prestige and the ability to inspire pious donations did confer a real advantage when it came to adopting a technology that depended on procuring access to land—it is one of the reasons that religious institutions were the largest class of conduit builders.

3 Design and Construction

In the year 1216, the monks at Waverley Abbey were facing a crisis. To their great consternation, the spring that had fed their aqueduct had dried up. Luckily for the abbey, one of their own, Brother Symon, put his mind to the problem. "With great difficulty, inquiry, and invention, and not without much labor and sweating," he was able to restore the abbey's supply by channeling dispersed veins of water together into a new "living and perpetual spring, made not by nature but by art." From the new artificial conduit-head, a dependable supply of water could be conveyed to the offices of the abbey.[1] Brother Symon was not alone, either in his difficulties or in his eventual success. Although they worked within inherited technological traditions, medieval engineers and craftsmen had to adapt whatever hydraulic experience and expertise they had to specific exigencies and circumstances. Few medieval water systems were constructed without a degree of difficulty, inquiry, and invention. All required labor and sweating.

In general, the best evidence for the construction and configuration of water systems comes from workmen's contracts, building accounts, and archaeology. In England these sources are supplemented by three medieval waterworks plans and a detailed narrative description of the construction of Waltham Abbey's conduit.[2] By bringing together these various types of evidence, it is possible to begin to reconstruct the steps involved in fabricating hydraulic components and in linking them together into functioning systems. In addition, one can catch occasional glimpses of the individual craftsmen and laborers who carried out the tasks and attempt to analyze how well the hydraulic systems, the end results of such hard labor, actually worked.

The construction of a complex water system required a high initial investment of labor and money and the organization of a workforce composed of both laborers and specialist craftsmen. Monks like Brother Symon played a part in some construction campaigns, but much, if not most, of the work seems usually to have been done by laymen. An enthusiastic description of the construction of the second abbey at Clairvaux paints a rosy (if somewhat imprecise) picture of bustling activity. "Resources flowed, workmen were swiftly assembled, and the brethren too threw themselves into the work in every way." Among their achievements were the diversion of the river and the channeling of its water into buried conduits, which carried the water to every building, as well as to the mills and other water-driven machinery, before returning it to the river. Although the description indicates that both monks and laymen worked on the construction of the new water system, it is not clear how the work was distributed or who had the necessary expertise. At Norton Priory, the construction of the thirteenth-century waterworks channels, moat, and drains would have required the removal of an estimated 1.6 million wheelbarrow loads of clay, an undertaking that would have taken forty laborers some three years to complete. Clearly this was a substantial and ambitious project: did the canons hire laymen to do the work, or did they do part of it themselves?[3]

Detailed building accounts, such as those from Exeter Cathedral, allow a more detailed look at the construction process, since they give the wages and fluctuating, week-by-week composition of work crews. At Exeter the main building campaign began in midsummer 1347 and lasted for two years, work continuing even as the Black Death raged through the city. Workmen were hired by the day or the week, and the numbers and occupations of those employed on the job at any one time depended on the labor requirements of the different stages of the work.[4] Here all the members of the conduit crew seem to have been lay workers, some engaged long-term and others hired for short periods as the need for their services dictated. Waltham Abbey's conduit, too, was constructed by a team of laymen.

In terms of their physical configurations, complex medieval water supply systems were composed of three basic subsystems: the first collected the water at the natural source, the second conveyed it to the desired destination, and the third distributed it to users. Each subsystem had several possible configurations and combinations of components. At the conduit-head, water was collected from springs or aquifers and conducted into collection tanks, reservoirs, or conduit houses. The conduit-head subsystem was often designed to perform a dual role: it served as a collection system and as an initial water purification system. From the conduit-head the water was conveyed to its final destination by pipes or channels, or both, and might flow through various intermediate tanks, filtering devices, water towers, or sluices en route. These subsidiary components regulated the pressure and flow and provided additional means of removing impurities. At the termination of the system, the water was distributed for collection or immediate use, usually by means of fountains or taps.

Drains often were used in conjunction with complex supply systems, but they are best seen as separate (albeit closely related) hydraulic systems rather than as mere subsystems of the main supply. Unlike the structurally linked supply subsystems, drainage systems could function independently and usually carried away waste products and runoff, as well as any excess water from the supply system itself.

CONDUIT-HEADS

The first component of a medieval water supply was the intake system. This could take various forms. Chamber intakes were the most common, but on occasion well-type intakes and adit intakes were employed. Where available, springs were the preferred sources for intake systems that supplied water for consumption or washing. The Romans had used both springs and rivers as sources of water for aqueducts, but they preferred the quality of springwater. Medieval Europeans seem to have held the same opinion. Although groundwater was more subject to chemical adulteration (as it picked up ions from the surrounding rocks and soil), it was far less subject to physical pollution than surface water.[5] Furthermore, in the dry

Mediterranean region, springs and aquifers provided more constant and reliable sources of water than seasonal surface supplies. Even when a spring-fed intake system was available, however, it was often supplemented by a secondary intake system of diverted river water, which could be used for purposes such as flushing drains, where water quantity was more important than its quality.

Spring-fed water systems normally had an initial intake reservoir, where water was collected and fed into the main pipe. In addition to collecting water for distribution and establishing the initial head of water, the reservoirs could serve as filtration and settling tanks. Intake structures varied in complexity. Some conduit-heads were simply open ponds or cisterns, some had more elaborate catchment pits and tanks, and some were protected by small buildings known as conduit houses. Many systems seem to have combined several of these features to create complex, composite conduit-heads.

The land transaction charters for aqueducts often include provisions for the construction of some sort of intake structure—the general type of structure and its approximate dimensions can sometimes be determined from the terms of the grant. The 1237 conduit grant from Gilbert de Sanford to the city of London included both springs and permission to construct a "tank (*castallum*) or pond (*piscinam*)" to collect the water. The Friars Minor of Exeter were given permission to enclose two small springs of water in the bottom of the city ditch with a low stone wall and to lead them by underground pipes to their house. The grant of a spring to Oseney Abbey near Oxford included the right to build a house eighteen feet long and thirteen feet wide over the spring and provided easements for repairing the building. In 1331 the Carmarthen Franciscans were granted a spring and permission to "erect a little house of stone, either round or square as they shall please, ten feet long and as many broad."[6]

In order to collect a sufficient quantity of water, the springs where intakes were constructed might have to be enlarged as well as enclosed, and dispersed natural veins might be artificially channeled together into the initial reservoir. Some grants contain explicit provisions for this phase of construction. The Oseney charter included permission to dig about the spring, whereas the Carmarthen grant included "liberty to dig and search for the veins of water of such spring, and to collect and conduct these by underground passages" to the conduit house.[7]

Ponds or open reservoirs appear at the heads of several medieval water systems. It is not always clear whether they were the sole components of a simple conduit-head or whether they fed more specialized structures. The building accounts for Vale Royal Abbey list the wages of "diggers and other common workmen working with trowels and hoes and other tools suitable for digging . . . upon the forming of a pond, from which a watercourse should flow down to the site of the monastery, and for making the mortar." The work crew consisted of nineteen diggers and one master, and the project seems to have been completed in a single week. The reservoir may have been simply an open tank—the payments for making mortar seem to indicate that its surface was at least lined. The original conduit-head for Saint Augustine's Canterbury seems to have had an initial open reservoir. Archaeological excavations have revealed the remains of a large clay-lined artificial catchment pond, which was replaced by a later conduit house.[8]

Some conduit-heads had multiple components. An early example of a composite conduit-head comes from Christ Church, Canterbury. The twelfth-century plans show two circular structures at the head of the water system. The first and second structures were linked by a pipe(?), and the water was then discharged through an outflow pipe(?) in the second structure. The first circle is thought to represent a simple catch-pit. The second may have been a more elaborate catch-pit (like the Waltham *piscina* described below), but it is called a "turris" on the plan. This could imply that it was a more substantial, standing structure, perhaps a round conduit house. It appears to have a perforated circular plate in the bottom of the basin, at the intake of the outflow pipe. There is no sign of an overflow pipe or a purge pipe.[9]

A plan of a composite conduit-head, along with a narrative account of the construction of the water system and a detailed description of the *placea capitis fontis* (site of the source of the spring) is preserved in the cartulary of Waltham Abbey (see fig. 3.1).[10] The narrative tells how, in 1220, Master Laurence of Stratford (a metalworker who seems to have acted as the chief hydraulic engineer and director of the conduit project) and a crew of about twenty workmen dug around a spring in Wormley that had been granted to the abbey. They reached the head of the spring and hit a good hard bottom within three days but found that the water issued forth in three separate fissures.

To collect the water, the Waltham team dug small trenches "in hard and

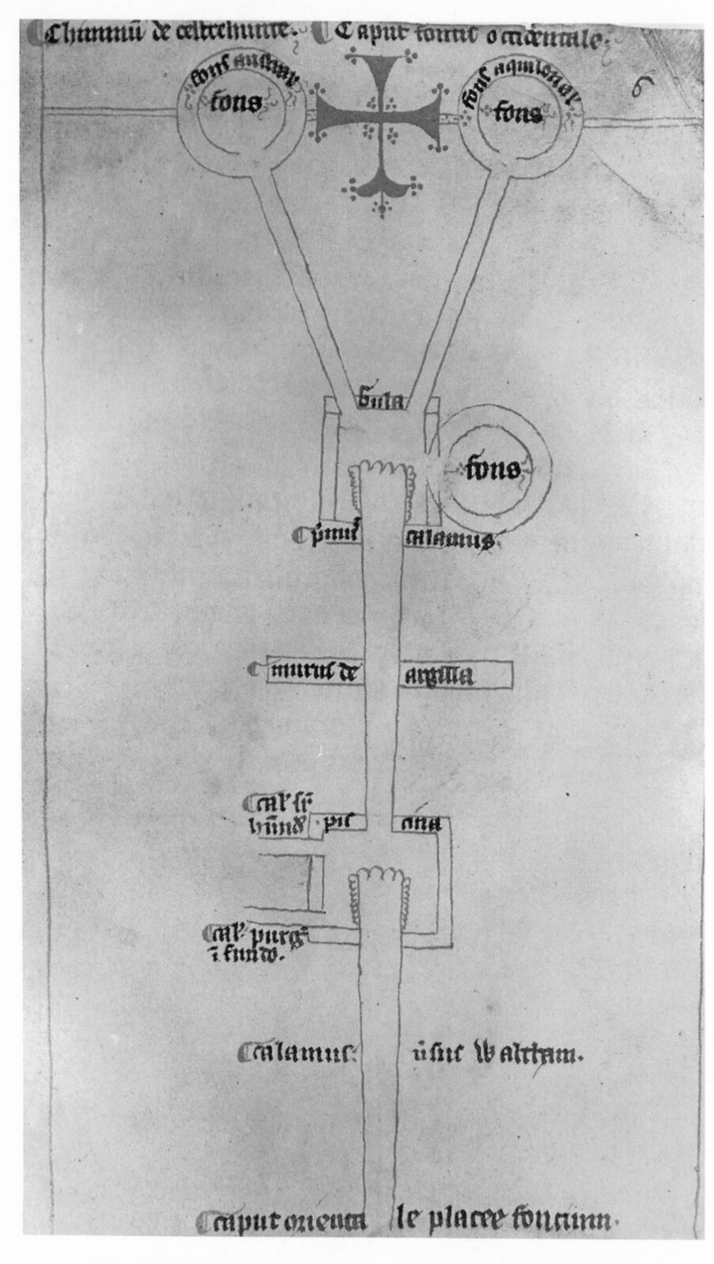

FIG. 3.1. Waltham Abbey waterworks plan. British Library Harley Ms. 391, fol. 6r. Reproduced in R. A. Skelton and P. D. A. Harvey, *Local Maps and Plans from Medieval England* (Oxford: Clarendon Press, 1986), pl. 2.

stony soil" that conducted the water from each natural head into a rectangular pit. The trenches were filled with washed flints so that the water would run between clean stones (presumably a deliberate attempt at filtration). The rectangular pit was also filled with washed stones; it was known as the *gula* (opening) and functioned as the initial collection and filtration reservoir. There is no indication that the gula was covered, and it seems to have been a simple pit rather than a masonry tank. From the gula the water entered the first pipe, which had a perforated intake to further filter out debris and which was situated at the bottom of the pit. The gula does not seem to have been provided with an overflow pipe, but a clay wall was built between it and the next tank to divert any floodwater.[11]

Ten feet beyond the gula and past the clay wall, the first pipe discharged into a small *piscina*. This was an oblong masonry tank "like an altar," lined with squared and polished freestone and covered with a single slab of marble. Water from the piscina discharged into three outflow pipes: the large main supply pipe to the abbey, a secondary supply pipe that fed any overflow to the court of the abbey's advocate (and donor of the springs), Henry of Wormley, and a purge pipe so that the tank could be emptied and cleaned. The outflow pipes were situated according to their function. The purge pipe (which had some sort of plug) discharged out the south side and was placed at the bottom of the tank so that the piscina could be completely emptied. The large orifice of the abbey supply pipe was located in the middle of the east side, about halfway up the wall of the tank, so that it received a full supply of water but was free of any sediments settling in the tank's bottom. The plan shows its end protruding into the piscina with a scalloped edge. The advocate's pipe was located near the top of the tank on the side, so that it received water only after the abbey's needs had been fully met. It is described as "curved like a bow" so that malicious persons could not pollute the water in the tank by poking filth back through the open pipe. The entire operation of digging for the springs; making the gula, the piscina, and the clay wall; and laying the associated pipes seem to have taken a little over a month, since the construction crew was at work on this section of the conduit from May 25 to early July.

Open collection basins at the conduit-head were vulnerable to the threat of pollution, whether by natural or human agency. In order to protect the purity of the water supply, and perhaps also to keep unauthorized persons from stealing water out of the tanks, many water systems were provided with small, roofed buildings over the reservoirs at the conduit-head.[12] Several of these medieval conduit houses are still standing (often with later modifications), and others are known from documentary sources. Although they have greater architectural elaboration, the basic hydraulic features are those already encountered: a tank or cistern large enough to hold a sufficient quantity of water, an intake pipe or channel, an outtake pipe, and perhaps an overflow mechanism and a drain or purge pipe so that the tank could be emptied for cleaning.

At Mount Grace Priory, three springs on the hillside were provided with their own individual conduit houses. Excavations of two of the buildings revealed enough of the structures for reconstructions to be made, although the lead tanks and pipes had been robbed. At Mount Grace, as in many cases, the tanks were below ground level. At Lichfield the conduit house was cut into a steep slope, with the cistern cut out of the living rock. The first conduit house of the London Franciscans' water system lay hidden four feet underground (the register gives instructions how to find it), whereas the upper parts of the second were visible from a distance. The subterranean cisterns of both buildings were approached by descending a flight of steps.[13]

Some conduit houses were components of composite collection systems. The fifteenth-century London Charterhouse plan shows a series of wells and springs connected by stone gutters (and in one case a lead pipe with a perforated end) to a "house of stone which receives the water of the same well and of the other wells and springs." At Wells the conduit house was part of a conduit-head system that also had "dikes, trenches, ponds, cisterns, etc." The conduit house for Saint Augustine's Abbey at Canterbury was fed by an adit-intake system composed of radiating tunnels. The tunnels fed water from twenty-four separate springs through stone-lined channels into the main structure.[14]

Most conduit houses were square or oblong, but some were round or polygonal. A thirteenth-century circular conduit house with a domed roof and shouldered doorway still exists at Beaulieu Abbey. Saint John's Well, one of the Mount Grace conduit houses, is a circular water tank with a pyramidal stone roof. The conduit house of Saint Augustine's Abbey at Canterbury is polygonal. At Wells a 1451 conduit grant from Bishop Beckington to the city gives detailed specifications for a circular house at the conduit-head: it was to be ten feet in internal diameter, with walls of stone,

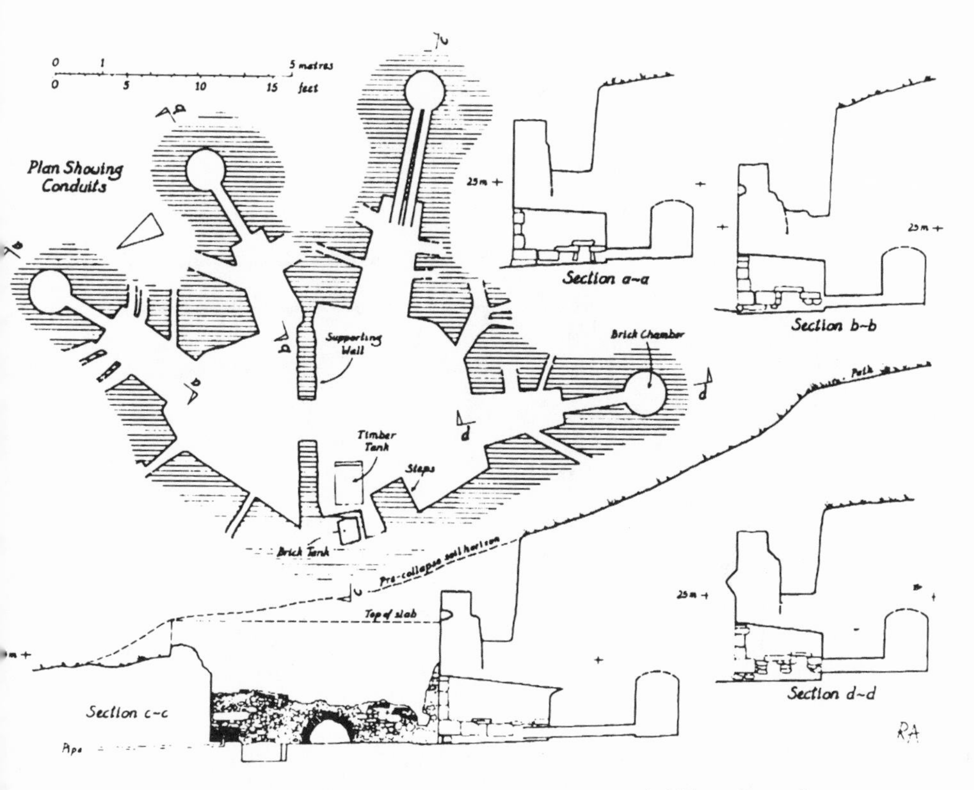

FIG. 3.2. Conduit house plan and sections, Saint Augustine's Abbey, Canterbury. Four main tunnels (with subsidiary ducts) and three smaller ducts fed water into the octagonal central reservoir. The adits tapped water from twenty-four springs. The medieval conduit house was refurbished in the eighteenth century. Paul Bennett, "St. Augustine's Conduit House," *Canterbury's Archaeology* (1987–88): 9. Reproduced by permission of the Canterbury Archaeological Trust.

lime, or other suitable material. Inside it was to hold a round lead cistern five feet deep and four feet in diameter, with pipes attached to either side. (Since the water was to be divided equally between the bishop and the city, presumably each had a supply pipe attached to the cistern.) The building was to have one door and two keys (one held by the bishop, one by the city).[15]

In some cases the entire interior of the conduit house served as the reservoir; other conduit houses had internal divisions between the cistern and a dry area. Bishop Beckington's specifications called for a smaller lead cistern within the Wells conduit house. The London Charterhouse plan

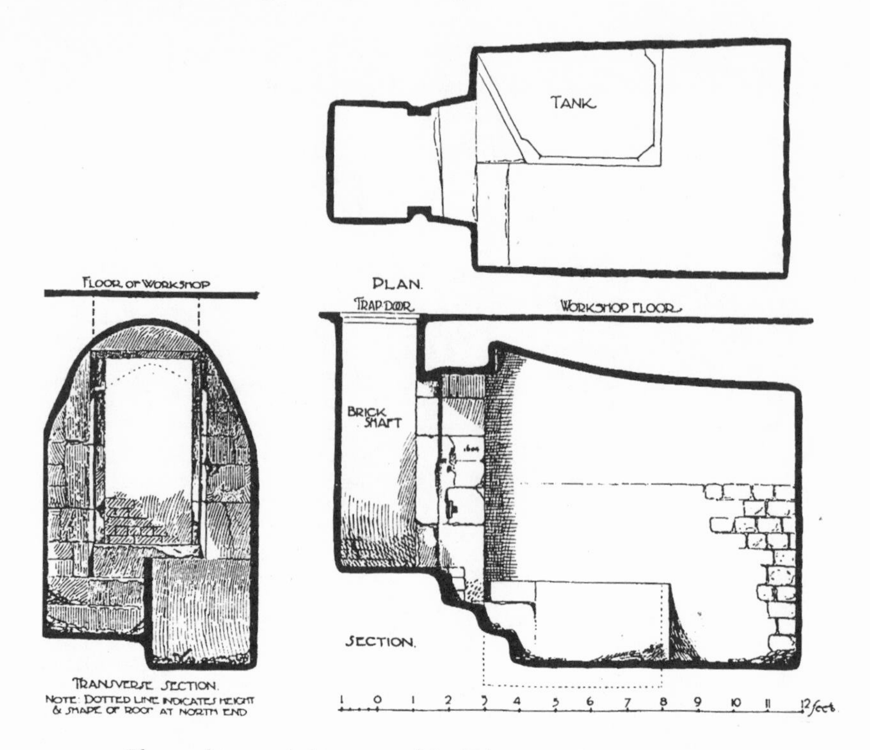

FIG. 3.3. Plan and internal elevations of the White Conduit, London. This was one of two conduit houses that supplied water to the Franciscan water system. After the suppression of the friary, the water system served Christ's Hospital. Philip Norman and Ernest A. Mann, "On the White Conduit, Chapel Street, Bloomsbury, and Its Connexion with the Grey Friars' Water System," *Archaeologia* 61 (1909).

shows several conduit houses. Some have separate, internal cisterns, whereas in others the reservoir completely fills the building.[16]

The overall dimensions of medieval conduit-heads are modest. The conduit grant to Oseney Abbey gave permission to build a conduit house with (presumably external) dimensions of 18 by 13 feet (5.5 × 4 m). The size of the cistern inside would naturally have been smaller. The Carmarthen grant permitted the friars to build a round or square building 10 feet (3 m) broad. The internal dimensions of the Waltham *piscina* are given in the narrative as about 5 by 3½ feet wide and 4 feet deep (1.5 × 1 × 1.2 m). The Lichfield conduit house cistern measures 1.74 by 1.07 meters (5 ft. 8½ in. × 3 ft. 6 in.) in width and is 2.21 meters (7 ft. 3 in.) deep. At Wells, the specifications called for a round lead cistern 5 feet (1.5 m) in diameter and 4 feet (1.2 m) deep, within a conduit house that had an internal diameter of 10 feet (3 m). The internal chambers of the two London Greyfriars' conduit houses measure some 9 by 6 feet (2.7 × 1.8 m) for the nearer head and 10 feet 6 inches (3.2 m) square for the more remote head.[17]

The capacities of these reservoirs are modest. In themselves, they do not provide enough data to calculate the overall amount of water available in their associated systems, but one is left with the undeniable impression that spring-fed medieval intake systems never supplied really large quantities of water.[18] They were designed to supply a limited quantity of high-quality water for a limited number of particular uses. Because of their restricted supply, spring-fed pipe systems were often complemented by river-fed open-channel systems, which could provide large quantities of lower-quality water for purposes that did not require a pure supply.

SEEPAGE TUNNELS

Seepage tunnels (filtration conduits) are special types of channel intakes. These are adits that penetrate the slopes of hillsides in order to follow the veins of subterranean aquifers and convey the water to a fountain on the side of the hill. The floor of the tunnel may have a cut and lined channel to carry the water, which percolates in from the sides and roof. Seepage tunnels were known in antiquity and are characteristic of some medieval Italian water systems. A particularly elaborate system of seepage tunnels, known as *bottini,* supplied Siena. In the tradition of antique tunnels, the Sienese bottini have vertical shafts (*smiragli*) at intervals along their lengths. These shafts could be used to fix the line of the tunnel. They also provided multiple work faces and points where spoil could be removed while the tunnel was being excavated, served as ventilation shafts, and provided access points for inspections and repairs. (They could also be dangerous: in 1341 Ambrogio da Piombino fell down a smiraglio and was given an indemnity of three days' pay.) Fountains were built on the spots where the bottini entered the hillsides.[19]

CONVEYANCE

The use of closed pipes for conduit systems was the prevailing practice in medieval England. Unlike the Romans, who generally employed open channels in their long-distance aqueducts and used pipes mainly to distribute water within cities, English engineers preferred to lay long runs of pipes from the water source to the final destination. In Italy, however, where antique aqueduct arcades were still visible, there seems to have been more of a tendency to follow the Roman tradition of long-distance channel conduits—indeed, some medieval Italian systems incorporated stretches of Roman aqueducts. At least some of the channel aqueducts seem to have fed local distribution systems of lead or earthenware pipes.

Medieval water pipes, like their Roman predecessors, were mainly made of lead, terra-cotta, and wood.[20] For the most part pipe systems depended on gravity flow, although pipes were also used for low-pressure systems, in which the pressure was determined by the head of water. Lead pipes were used for intake systems, waste pipes, and downspouts. In England it was common to use subterranean lead pipes for long-distance conduits—these carried water from the source at the springs for distances of up to several miles. Ceramic pipes were also used for intake systems and drains, although they do not seem to have been as common as lead pipes. Wood pipes were used both for intake systems and drains. Several sites have produced evidence of more than one type of pipe.

In medieval water systems, channels sometimes formed part of the main intake system. In many cases, however, channels were parts of two- or three-tiered water systems. A restricted quantity of potable water for consumption and washing was supplied by pipes or wells, or both, whereas channels delivered larger quantities of subpotable water for other uses: supplying fishponds, driving mills, other industrial uses, and irrigation. Channels were also widely used for drains and sewers. Even sites with piped intake systems normally had a network of channel drains for the disposal of wastewater, storm runoff, sewage, and other refuse. Channels could be lined with wattle, timber, or masonry and were often covered. Even covered channels, however, did not run full; they operated by gravity flow. For the sake of hydraulic analysis, they can be classified as open channels.[21]

LEAD PIPES

Lead pipes were generally made by plumbers, craftsmen who had a specialist's familiarity with the material even if they had no particular expertise in hydraulic engineering. When the city of Bristol needed new lead pipes for the city conduits in 1376, it contracted with plumber Hugh White to make them to the required specifications. The plumbers at Exeter seem to have played a subordinate and limited role in the construction of the cathedral conduit. They are listed only intermittently in the weekly wage accounts, and though they were responsible for making (and probably joining) the pipes, they do not seem to have exercised a supervisory role over the masons and laborers who were constructing the pipe trench. In some weeks two plumbers (each assisted by his "boy" or "servant") were employed. Some plumbers, like others in the building trades, may have been semi-itinerant craftsmen, but at least one of the Exeter plumbers had been associated with the cathedral works for many years. "R. the plumber," who together with his son made lead pipes for the conduit in 1347, had been employed by the Cathedral since at least 1328; his main responsibility would have been the cathedral's lead roofs and gutters, but when the need arose, he could turn his hand to making water pipes. The London plumbers' ordinances, drawn up in 1365, show that making pipes was one of many jobs a plumber could expect to do. The ordinances regulate prices for the standard products of the trade. The price for working a clove of lead for conduit pipes was fixed at one penny (the same rate as working a clove for furnaces, "tappetroughes," and belfries) and twice the rate for working a clove for gutters and roofs, which suggests that extra care was needed in working the lead for some types of jobs, including the fabrication of water pipes.[22]

A broader role was played by Master Laurence of Stratford, who built Waltham Abbey's water conduit in 1220–22. Laurence was a skilled metalsmith, who worked in gold, silver, bronze, iron, and tin as well as lead. Laurence made the pipes and laver and seems to have served as the chief engineer and director of the works, supervising the workmen as they dug trenches, laid pipes, and built tanks at the springs. In his absence, work on the project came to a halt. In most cases, however, we do not know who made the pipes or what overall responsibility they had for the design or construction of the water system. The limited evidence available suggests variability rather than uniformity. The fabricators of a lead pipe could be specialist plumbers or more generalized metalsmiths, they might be local or nonlocal craftsmen, and they might (or might not) take responsibility for more than simply making and joining the pipes.

The author of the Waltham narrative does not indicate where the lead came from, but to judge from contemporary building accounts elsewhere, it would have been purchased either in a local town or at a market close to the mines (such as Saint Boltoph's Fair in Boston) and transported to the site by ship and cart. It seems that at Waltham the canons were responsible for providing the materials and equipment: when Master Laurence first arrived, the project had to be postponed because "we did not have the tools and other necessaries ready and on the spot." (While waiting for the supplies to arrive, Laurence made the canons a conduit for their ale.) Exeter, in contrast, already had a plumbery (along with other craftsmen's lodges not far from the cathedral in the zone known as Kalenderhay), so that plumbing tools and supplies of lead were readily available. When "R. the plumber" made five conduit pipes in 1347, he was able to fabricate them "from our own lead." The construction and maintenance of lead roofs and gutters would have been the main reason for the existence of a fully equipped plumbery, but its availability would have made it considerably easier to manufacture lead pipes for a water system. The ad hoc provisions at Waltham do not seem to have been very satisfactory. After making one batch of pipes at the abbey, Laurence returned home on four subsequent occasions to make new pipes.[23]

The precise chronology of the Waltham narrative allows one to calculate Master Laurence's productivity. Laurence and his sons began casting lead for the pipes on March 6, 1220, and had completed some two hundred pipes before Whitsunday (May 17). Given days off for holidays or breaks in the work because of bad weather or the need to obtain fresh materials, the figures suggest that they were making four pipes (i.e., casting two sheets) a day. Several aspects of the manufacturing process can be reconstructed from the narrative, and physical evidence from surviving pipes and archaeological sites can supply additional details. Once the necessary supplies had been procured, Laurence, assisted by his sons Ralph and William, melted the lead and cast it into sheets. The narrative does not describe the melting process, but a simple lead furnace, uncovered at Waltham in the course of recent archaeological excavations, may have been associated with the water conduit: it consisted of a hole in the ground in which lead was melted with charcoal. At Exeter a lead furnace was purchased for the conduit works, and a second furnace may have been purchased for the same purpose. Other equipment used in lead founding at Exeter included an iron trough, two *flotis* (meaning uncertain), and a new set of bellows made from timber, grease, and the hide of a horse dead of murrain.[24]

Analysis of the composition of medieval lead pipes from other sites indicates that the lead used for water pipes was generally very pure: it may contain traces of other elements derived from the original ore, but nothing seems to have been deliberately added. In some cases the lead seems to have been desilverized, and on occasion Roman lead may have been recycled. Many building accounts include payments for washing and founding lead ash (*cinis plumbi*). This presumably refers to the practice of resmelting litharge, a residual product of the silver industry's extraction of silver from argentiferous lead ores. An oxide of lead, litharge was reduced by heating it with carbon (charcoal or coal) in a small blast furnace. The process produced desilverized lead, but this was probably not a deliberate attempt to control the composition of lead used for water pipes. Lead left lying around the building site (or on nearby structures) was a tempting target for workmen or their assistants seeking a little extra income. The London plumbers' ordinances warn members of the guild against purchasing pilfered lead: "None shall buy stripped lead from the assistants of tilers, bricklayers, masons, or women, who cannot find warranty for it."[25]

The Waltham narrative does not describe the casting bed or table. It may have been a bed of sand (or a sand-covered board) with clay-lined wooden sides, designed to yield a lead sheet of the desired dimensions; at Waltham the sheets measured 11, 12, and 13 feet long by 14 inches wide. The author does not mention any cleaning of the surface of the cast sheet or flattening it by hammering, steps that have been suggested in the manufacturing process for Roman pipes. Each of the Waltham cast sheets was cut lengthwise, yielding two long strips, each 7 inches in width.[26]

Laurence next "bent them [the lead strips] in the form of a tube and made pipes." This step probably involved rolling the sheet around a wooden mandrel, a process that would have deformed and refined the original crystal structure of the cast lead. Medieval pipes with round, oval, and pear-shaped cross sections are known. The variations in shape have not been systematically studied, but they probably reflect slight differences in manufacturing techniques. To make the longitudinal seams, Laurence (after removing the mandrel) filled the pipes with sand, encased them in clay, and consolidated them with very hot molten lead. The clay would have formed the sides of a casting mold for the longitudinal weld that joined the edges of the rolled sheet. The high temperature of the molten lead poured into the trough would remelt the edges of the sheet and fuse them with a cast join. The process leaves a prominent ridge along the seam. Such ridges may have a large-grained columnar crystal structure (in contrast to the smaller grain size of the body of the pipe), which renders them vulnerable to intergranular corrosion along the crystal boundaries.[27]

Uniting the edges of the pipe by means of molten lead (a process known as autogenous soldering) was a common (although not the only) method used in Roman pipes and was the method used in several medieval pipes that have been subjected to metallurgical analysis. A preference for the use of autogenous solder rather than a lead-tin alloy to consolidate lead water pipes is mentioned by the thirteenth-century medieval encyclopedists Thomas of Cantimprè and Vincent of Beauvais, who considered it a recent innovation that rendered the pipes less vulnerable to corrosion when buried underground. The belief by these authors that the technique was a new one *may* indicate that the Roman art of autogenous soldering had been lost and subsequently reinvented.[28]

The adoption of autogenous soldering for medieval water pipes was not universal, however. A lead pipe from Roche Abbey was soldered together with a lead and tin alloy. In 1447 Henry VI granted the inhabitants of Westminster the right to convey the overflow from one of the Westminster Palace conduits to the town by means of a lead pipe, and the clerk of the King's Works subsequently sold them 4,857 pounds of lead and 46 pounds of solder for the project. The purchase of tin and beef tallow "for the pipes" in the 1348–49 Exeter Cathedral fabric accounts seems to indicate that the pipes were being joined with lead-tin solder (the tallow serving as a flux), although the metallurgical examination of an undated (possibly medieval) pipe from Exeter revealed a brazed lead seam. Salt, another flux, was used "for soldering (*consolidand*) the junctions on the pipe" at Leeds Castle in 1381 and "ad sowderandam [to solder]" lead conduit pipes at Gloucester in the late fifteenth century. The slender evidence available suggests that there may have been a shift toward the use of autogenous solder in the thirteenth century and a revival in the use of tin-based solders in the late fourteenth and fifteenth centuries, but any satisfactory determination of the temporal and geographical parameters of these different methods of pipe consolidation would require the metallurgical analysis of a much larger sample of securely dated pipes. In any case, the use of autogenous solder for subterranean water pipes does not seem to have been part of any widespread rejection of tin-based solders, which continued to be used for soldering lead gutters (and possibly downspouts), roofs, and window cames.[29]

Although the overall manufacturing method described in the Waltham narrative seems broadly consistent with the physical evidence from surviving pipes, there may have been variations in some steps of the process, such as making the pipe seam. The cross sections of ridges along the seam in medieval (and Roman) pipes come in several forms. One common type is distinctly angular and takes the form of an inverted V; others form a more rounded inverted U. The macroscopic morphological differences apparently reflect variations in the manufacturing process. Tylecote, examining Roman pipes, suggests that the inverted V seam is formed by slightly hammering the weld after it is cast. Coppack has suggested a different method for making these angular seams: a V-shaped channel is made by impressing a wooden template in a casting bed of sand. The rolled sheet of lead is placed with the open seam on the bottom, so that it rests on top of the depression, and molten lead is poured into the channel, thus forming a ridge with an inverted V cross section. Desch, who examined a "blobby type" (inverted U) pipe seam from Exeter, concluded that the seam had been made of molten lead and then finished with a burning iron.[30]

The Waltham strips of lead, 7 inches (17.8 cm) wide, would have yielded pipes approximately 2¼ inches (5.7 cm) in diameter (assuming that the edges were butted together when the seam was made). Cross sections of medieval pipes often show a slight indentation on the interior surface at the midline of the join, indicating that the edges of the lead sheet were brought in direct contact before the join was fabricated. The capacity of the first batch of pipes was found to be inadequate when the water source proved more abundant than anticipated, so Master Laurence "enlarged his pipes by one inch"—that is, he increased the width of his lead strips to 8 inches (20.3 cm) (and hence increased the circumference of his pipes).[31] These larger pipes would have had diameters of about 2½ inches (6.35 cm).

The dimensions of the pipes given in the Waltham narrative are somewhat larger than an actual pipe found on the abbey site (which had an external circumference of slightly less than 6 inches) but fall well within the range of medieval lead pipe dimensions known from other sites. Publication of lead pipe dimensions in archaeological reports needs to become more standardized, but generally pipes known from English sites range from about 1 to 4 inches (2.5–10 cm) in external diameter and ¾ inch to

3 inches (2–7.5 cm) in internal diameter. Medieval documents sometimes refer to lead pipes with the diameter of a goose or swan's quill or a little finger. These small diameters are usually specified in grants for branch pipes from existing systems and are clearly intended to limit the amount of water removed from the main supply. The specified diameters may refer only to the narrow termination of the pipe. A segment of brass pipe attached to a late-fourteenth-century Bristol memorandum demonstrates that, in one case at least, the swan's-quill diameter was considered to be .3 inch (7.6 mm).[32]

The Waltham narrative suggests that Master Laurence was standardizing his pipes' diameters by making his lead sheets an exact number of inches wide. The widths chosen for the sheets, however, were more variable than the standardized sizes prescribed by Roman authors. Seven inches, the width of Laurence's lead strips, was not a Roman pipe dimension. The purpose of Roman standardized dimensions had been to regulate water consumption among paying customers; since medieval pipe systems were not organized on this basis, there is no reason why they should have adhered to a single system of normative dimensions. Nevertheless, standardizing the width of the lead sheets within a particular pipeline would have been an obvious means of ensuring that the pipes would fit together.[33]

According to Vitruvius, the prescribed length for Roman pipes was ten feet. Medieval craftsmen do not seem to have been bound by a standardized length, although lead pipes of about ten to twelve feet in length seem to have been the general rule. This was probably due to practical rather than theoretical considerations, since it would have been inconvenient to work with pipe segments that were considerably longer or shorter. Short pipes required too many joints; very long pipes would be extremely heavy and cumbersome. A Bristol contract of 1376, however, specifies that plumber Hugh White is to make new pipes for the conduit, "twenty feet in length and not more."[34]

After the individual pipe sections were finished, they had to be joined together. Pipe joints were vulnerable and were likely to leak if not carefully made. At present too few examples are known (and the published examples are too unevenly described) to establish a fully satisfactory typology of medieval pipe joints. Several methods seem to have been used. The Dover Castle pipes had simple butt-joints, which were sealed and reinforced by pouring molten lead over them. The lead was roughly wiped. The joints of the Kirkstall Abbey pipes appear to have been reinforced in a similar fashion. Some of the Kirkstall pipes have simple butt-joints like the Dover pipe, but others were joined by inserting the tapered end of one pipe into the slightly flared end of another. It has been suggested that the raised collars over the pipe joints at Saint Alban's Abbey were produced by pouring molten lead into clay molds. Some pipe sections were joined by soldering on precast jointing collars. An example from Meaux Abbey has a flared-tapered soldered join, with a strip of lead soldered around the outside.[35]

The account rolls from Durham Abbey record numerous purchases of tow, cord, spun yarn, linen cloth, canvas, and hemp for "binding" the aqueduct. The precise method in which these materials were used is not clear. They may have served to strengthen the joints by lashing the pipes together, a technique that has been suggested for some large Roman pipes. Tallow, kitchen drippings, pitch, and resin were sometimes used with the cords and cloth. Bindings saturated with such unctuous materials may have been used to mend leaks, to make packed gaskets at the joints, or to help insulate the pipes to prevent them from freezing. A sixteenth-century record of expenses indicates that such materials could also be used as a substitute for soldering the joints when necessary: "clothe and tallowe to bynd [the conduit] in some places wher yt could not be sowdered." Work on the pipes was a frequently recurring expense for Durham, and both men and women were employed in the fight to keep the system running.[36]

Fabrication of pipe joints must often have taken place close to the final position of the pipe or even after the pipe had been placed in the trench. Individual sections would be easier to carry than jointed lengths of pipe; furthermore, joints made on the spot would not be subject to the jolts and stresses of transport. The employment of teams of plumbers at Exeter during the weeks the pipe was laid suggests that at least some of them were making joints. At Waltham Abbey, Master Laurence and his sons seem to been closely involved with the laying of the pipe: as well as supervising the excavation of the pipe trench, they were probably making the joints as the work proceeded. The plano-convex cross section of the boss surrounding one Kirkstall Abbey joint indicates that it was made after the pipe had been laid in the ground. Occasionally, joints are marked. The number "IV" was chiseled into the upper surface of a Meaux Abbey joint, and a bossed joint from Waverley Abbey had a diagonal cross marked on top. The function of these marks is unknown. Coppack does not think the Meaux Abbey "IV" is a plumber's mark, but he suggests that it may indicate the pipe's function in the plumbing system. If the marks were intended to assist the workmen in laying the pipes in the correct position, the casting of the joints must have taken place at some distance from the trench. Alternatively, they may have served as guides for future maintenance. In either case it is tempting to speculate that such marks may have been keyed to (lost) water-system plans.[37]

The full range and distribution of possible variants in medieval lead pipe manufacturing and joining methods have yet to be established. Further progress on these questions would be greatly facilitated by a detailed metallurgical investigation of a larger sample of medieval lead pipes of known date and provenance.

EARTHENWARE PIPES

The most common medieval alternative to lead pipes was ceramic pipes. Almost all the evidence for medieval earthenware pipes comes from archaeology. There is no documentary source equivalent to the Waltham narrative or the Canterbury plan. Three main types of earthenware pipes are known from medieval contexts. The simplest is a plain, tapered tube. These conical pipes are joined by inserting the narrow end of one pipe into the wide end of the next pipe. The second type has a flange at the narrow end and is slightly flared at the wide end: the narrow end is inserted as far as the flange into the splayed end of the adjoining pipe. The third type has a shoulder (but no projecting flange) at one end. The shouldered end is inserted into the plain end of the adjacent pipe, which may be either flared or cut away on the interior to form a socket.[38]

The significance of these different types is not immediately apparent. It does not appear to be chronological: both plain and flanged pipes have been found in twelfth-century contexts in Britain. The date of the earliest shouldered pipes has yet to be determined (and may be slightly later than the other two types), but all three types remained in use in the later Middle Ages and beyond. Nor do the variant types appear to reflect regional differences—in fact, some sites have produced more than one type.[39]

All three types of earthenware pipes were coupled by overlapping joints. Greek and Roman earthenware pipe joints had been sealed with an expanding paste, made from a mixture of quicklime and oil. Mortar does seem to have been used to join some medieval earthenware pipes—the Ely pipes retained traces of white mortar at the joint—but in other cases the join was sealed with clay or possibly even left unsealed. This may have been an attempt to reduce pipe fractures by keeping the joint flexible: the use of cement mortars renders the joint (and hence the entire system) rigid, so that the joined pipes act as a beam and become vulnerable to breakage as a result of soil movements.[40]

The dimensions of earthenware pipes were limited by the manufacturing process. Medieval wheel-thrown pipes generally ranged between twelve and twenty inches (about 30–50 cm) in length. The short lengths of pipe meant that an earthenware pipeline had many more joints than a lead or wooden one. Keeping the lengths of a terra-cotta pipe short, however, reduced the chances of the pipe's fracturing once it was laid. The minimum internal bores of medieval earthenware pipes range from about one and one-half to four inches (about 4–10 cm). These figures are generally comparable to medieval lead pipes, although they exceed lead pipe diameters at the upper end of the range.[41]

Although earthenware pipes could be (and sometimes were) used for pressure systems, their frequent joints, the difficulty of achieving a tight seal, and their low tensile strength rendered them less suitable for this purpose than lead or wood pipes. Many ancient and medieval earthenware pipe systems were designed to flow only partly full—their hydraulics were those of open channels rather than closed pipes.

The system of overlapping joints used to couple medieval earthenware pipes affected their hydraulic efficiency. With the exception of some shoulder and socket pipes, the bore of medieval earthenware pipelines was not uniform. Instead, it regularly contracted and widened with each length of pipe. Since velocity of flow is a function of a bore's cross-sectional area, water flowing through such a system would be alternately accelerating and decelerating, giving rise to eddying and resulting in an overall retardation of the rate of flow and discharge. The hydraulic efficiency of earthenware pipes was also impaired by the roughness of the internal surfaces (the ridges produced by throwing on a wheel would be of particular consequence) and the inherent inaccuracy in the alignment of the joints, where lipping could cause blockages. Such effects may not have been particularly serious in intake systems, when the water was relatively clean, but they could cause problems when earthenware pipes were used for drainage: it is no accident that the Thetford earthenware drain pipes retained a "heavy deposit of silt" on their inside surfaces.[42]

Earthenware pipes were a specialized product of the medieval pottery industry. Some pipes were formed by hand, but most seem to have been thrown on a wheel in the manner of a pottery vessel, a process that leaves spiral ridges on the interior surfaces. After the walls had been formed, the "base" was cut out with a knife. Exterior surfaces were smoothed, or might be knife-trimmed longitudinally. Flanges were made by either pressing out the vessel wall (presumably while the pipe was still on the wheel) or adding a separate flange to the body of the pipe. At Canterbury, the flange junctions of the pipes were smoothed and then stabbed diagonally through both flange and body, perhaps as a means of facilitating even drying to reduce losses during firing.[43]

Some pipes were also glazed. Glazing could be, as at Thetford, merely the result of unintentional splashes, but other pipes were deliberately glazed, either overall or at the ends. The purpose of glazing pipes is not clear. Glazing the interior would improve the hydraulic efficiency by reducing surface roughness, but the Ely pipes, for example, are glazed only on the exterior surface and at the ends. Pipes from Reigate were glazed only within the narrow end: this may have had something to do with jointing, or just possibly it helped workmen in the identification or orientation of a particular batch of pipes. It is possible, however, that a glaze was sometimes added simply because that was the customary method of treating other ceramic products at the kiln site.[44]

The manufacture of earthenware pipes seems to have taken place at potteries with a wide range of products. Few British kiln sites have, as yet, produced direct evidence for pipe making, and the process may have been too specialized for the ordinary potter. Potters who made other building components, such as roof fittings, are the most likely candidates for water pipe manufacturers. In particular, the basic morphological similarity between earthenware pipes and chimney pots meant that the technical expertise gained in forming and firing the one form could be readily transferred to the manufacture of the other—as could business expertise in marketing building components.[45]

There is some archaeological evidence to support the hypothesis that earthenware water pipes were made at potteries that produced unusually wide ranges of ceramic products, including other building components. At Ely, water pipes exhibited the same fabric and glaze as roof fittings found nearby. The technique of piercing the vessel with stab-holes, noted on pipes from North Lane, Canterbury, was also used on chimney pots found at the same site. The most direct evidence comes from the pottery kilns at Laverstock in Wiltshire. The Laverstock potters made an exceptionally wide range of products, including many types of roof fittings, and water pipes have been found at several of the Laverstock kilns. Pipes have also been found on medieval kiln sites at Tyler Hill in Kent and at Bourne in Lincolnshire; in both cases kiln products included other building materials as well as domestic vessels.[46]

Many pipes seem to have been of fairly local manufacture, though whether they were only produced to order for individual projects is unclear. One would expect that "mass-produced" pipes intended for large jobs or commercial distribution, such as the seven thousand pipes made by the potter Ulrich for the town of Freiburg in 1501, would be standardized and interchangeable. This may be the case, but the regrettable tendency of archaeologists to publish drawings of only one or two "representative" pipes from larger collections makes it difficult to estimate the degree to which batches of pipes were standardized.[47]

Whether earthenware pipes, like tiles, were also made by itinerant craftsmen in kilns close to the building site is another question that remains unanswered. Some systems, such as the one at Glenluce Abbey, suggest that the manufacture of pipes may have taken place close to the site. The Glenluce pipes were provided with tally marks scored in the wet clay: these apparently were intended to ensure tight jointing by indicating which specific pipes and junction boxes were designed to fit together. Such concern with individual components indicates that the Glenluce pipes were made for a particular water system; if so, there could have been some advantage in making them close to the site itself, where direct measurements for the various pipe runs could be taken.[48]

Medieval plumbers both manufactured and installed lead pipes; some, like Master Laurence at Waltham, seem also to have acted as hydraulic engineers and project directors. Did the potters who made earthenware pipes also take part in the installation of their products? In most cases they probably did not. The jointing of earthenware pipes with clay or mortar did not require the specialized skill needed to couple lead pipes, so it could have been performed by ordinary masons or general workmen (although an experienced engineer would still have been needed to design and oversee the construction of the system as a whole). Earthenware pipes found in a stack at Griff Manor House seem never to have been laid: perhaps after purchasing the pipes, the owners were unable to find someone to install their water system.[49] If pipes were made at the building site, however, the potter may have played a role in their installation. The Glenluce pipe maker seems to have been concerned with fitting individual pipes into an overall design, and it is possible that he did supervise their installation.

WOOD PIPES

Pipes made of wood, a cheap and readily available material, were very common in northern Europe during the Roman and postmedieval periods. In Germany and Switzerland, wood was the preferred material for medieval urban pipes, whereas monastic pipes were usually made of lead, or occasionally earthenware. Wood was certainly used for water pipes and drainpipes during the Middle Ages in Britain, but it does not seem to have been as popular a pipe material as lead or terra-cotta. The apparent scarcity of wooden pipes in medieval British contexts may arise in part from the accidents of survival: an organic material, wood is infrequently preserved in archaeological contexts. Some wooden pipes are known only from the voids they left in the surrounding soil; the traces left by others may have eluded the excavators. Nonetheless, the German pattern of lead monastic pipes and wooden urban pipes does not seem to fit the British evidence very well: here, urban systems were more likely to use lead or earthenware pipes (at least in the medieval period), whereas the known examples of wooden pipes are as often as not associated with religious houses.[50]

Wooden pipes were made either by hollowing out tree trunks and closing them with a plank or by boring them longitudinally, probably with a long auger. The former method resulted in pipes with variable internal dimensions. As the diameter of the trunk narrowed, so did the internal diameter of the pipe. The latter method produced a regular bore and was more suitable for pressure systems. Oak was commonly used for Roman wooden pipes and has been identified as the type of wood used for pipes at Thornholme Priory and Dublin. According to the terms of a 1369 contract, the water pipes for a range of London shops were to be made of oak. Elm was also used for water pipes, as at Beaulieu Abbey, where hollowed elm

FIG. 3.4. Boring logs and making them into pipes. Woodcut from 1556 edition of Agricola's *De Re Metallica.* Georgius Agricola, *De Re Metallica,* trans. Herbert Clark Hoover and Lou Henry Hoover (New York: Dover Publications, 1950), 177.

trunks have been found. The exterior of the trunk could be left in its original rough shape: the voids left by the decayed pipes at Aldermanbury, London, and at the Manor of the More were oval or round. At Thornholme Priory, however, a surviving pipe was square sectioned with a round bore.[51]

There is insufficient data to make generalizations about the dimensions of medieval wooden pipes. (The dimensions of voids can suggest the overall circumference of the pipe but sheds little light on the hydraulically significant diameter of the bore.) The Thornholme Priory pipe was drilled with a bore diameter of 4½ inches (11 cm), but it is difficult to say how typical this was. It is somewhat larger than the largest earthenware pipe bores (about 4 in./10 cm) and is about 1½ inches (4 cm) larger than the bores of the large lead pipes. The lengths of individual pipes would have been determined by the length of the tree trunk, and the long individual sections would have had the advantage of reducing the cost of labor on joints. At least some medieval wooden pipes were coupled in the manner used for Roman pipes, with a circular iron collar reinforcing the joint.[52]

Wooden pipes were made (and, it seems, installed) by carpenters. The 1369 contract mentioned above called for two London carpenters to build a range of twenty shops and included oak water pipes among the detailed specifications. The wood pipes associated with medieval fishponds were also made by carpenters. As in the case of lead pipes, both the fabrication and the installation of wooden pipes seem to have been the responsibility of the craftsman who specialized in the material used. Unlike lead pipes, wood pipes would have been relatively easy to transport, handle, and install.[53]

The durability of wooden pipes varied considerably, depending on the type of wood, its saturation, and soil conditions. Wood absorbs water, and an untreated wooden pipe can become completely saturated if the pipeline runs continually full and under saturation pressure. In its fully saturated state it can withstand decay indefinitely; according to modern specifications, this can be achieved if a head of twenty-five or more feet is maintained. In medieval gravity-flow systems, full saturation pressure throughout the length of a pipeline would be unlikely, although the advent of lifting devices and water towers would improve the chances of maintaining saturation pressure. Joints were particularly vulnerable to decay. According to German and Swiss evidence, typical working life spans of wooden pipes could range between ten and forty years, although some medieval wood pipes are still well preserved.[54]

PIPE TRENCHES

Bristol's agreement with plumber Hugh White called on him not only to make new pipes but also to put them safely in the ground and to cover them properly. Most medieval pipes were laid in subterranean trenches. The trenches maintained the gradient of the pipe and, if well designed, could minimize pipe failures resulting from corrosion, frost, and other forms of mechanical damage. The depth of a pipe trench can affect the probability of pipe fractures. Pipes in very shallow trenches are likely to freeze or to fracture under local point loads; and if they run under streets (as many medieval pipes did), they are vulnerable to fractures caused by surface vibrations and wheel loads. Deeply buried pipes may fracture under the pressure of the overload of earth. Unless the pipe is provided with some sort of external protection, intermediate depths are safest.[55]

The depths of medieval pipe trenches seem to have varied considerably. A late-twelfth-century pipe trench for the cloister laver at Durham was only 3 to 4 inches (7.6–10 cm) deep. The pipe for the rebuilt laver was laid in a trench that was deeper (9 in./23 cm) but still relatively shallow: if the cloister pipe trenches are representative of the system as a whole, it is not hard to see why Durham had recurring problems with fractured and frozen pipes. Most typical seem to be trenches a few feet deep, regardless of the type of pipe used. The earthenware pipes at Glenluce Abbey were about 2 feet (0.6 m) below ground level, and the trench for the lead pipe at Lichfield Cathedral seems to have been dug to a similar depth. The trench for the earthenware pipes at Saint Augustine's Abbey was 75 centimeters (about 2.5 ft.) deep—the minimum depth recommended for modern British supply pipes. According to the Waltham narrative, the lead pipes were laid 3 feet (about 90 cm) below the surface, except when they were carried across an intervening river in a 2-foot-deep (0.6 m) trench.[56]

Some medieval pipe trenches are known to have had very deep segments. In London, diggers down in a deep pipe trench were attempting to clear sediment out of a clogged conduit pipe when they were overcome by foul air. The London Greyfriars' pipe seems to have been laid at depths varying between 3 and 18 feet (0.9–5.5 m). In the late twelfth century, workman William of Gloucester was buried while laying a lead pipe at Churchdown, when the sides of the trench caved in. At the point where the accident occurred, the trench was said to be 24 feet (7 m) deep. Luckily for William, he was dug out alive, a miracle that he attributed to his prayers to the recently martyred Thomas Becket and which is commemorated in one of the stained glass windows in Canterbury Cathedral.[57]

Deep trenches were probably used to maintain the gradient of the pipe. A water system has two fixed points—the water source and the destination—which determine the maximum average gradient of the pipes or channels. A lower average gradient can be achieved by taking a less direct route, and the overall system may have a uniform slope or an irregular profile. At Exeter, a twelfth-century(?) pipe trench excavated at King William Street reached a depth of 3.25 meters (10 ft. 8 in.) in order to maintain a steady gradient as it crossed a ridge. Such care was not necessary: closed-

FIG. 3.5. Men digging trenches at a building site. Thomas Aquinas, "De regimine principium" (c. 1350), Rome, Vaticana MS Chig. M. VIII, 158, fol. 12. Reproduced in Bernhard Degenhart and Annegrit Schmitt, eds., *Corpus der Italienischen Zeichnungen, 1300–1450*, vol. 2, no. 3 (Berlin: Gebr. Mann Verlag, 1980), pl. 13d.

pipe systems do not need to maintain a steady gradient, as long as the pipe does not have air locks and as long as the point of discharge is lower than the intake. Later Exeter conduit builders showed a greater awareness of (or confidence in) the tolerances permitted by closed-pipe systems: the fourteenth-century pipes rose and fell with the local topography.[58]

Medieval gradients were probably established by using relatively simple surveying techniques. Instruments such as the "miner's level" (an A-frame with a plumb line suspended from the apex, which was used by masons and is mentioned in late medieval Sienese *bottini* tool inventories) may have been employed. When Cistercians were selecting a site at Øm (Denmark), one of the monks, Martin (praised as ingenious by the Øm chronicler), used a plumb line to establish that water in one nearby lake was higher than that in another, so that if the monks dug a channel between the two lakes, they could have running water. There is no evidence to suggest that more complex antique surveying and leveling devices, such as the *chorobates,* were used in the Middle Ages, although Roman texts describing surveying, such as Vitruvius and the *Corpus Agrimensores,* were pre-

FIG. 3.6. Mariano Taccola, drawing of a fountain and its source. The A-frame miner's level indicates that the water in the fountain will rise to the height of the head when the pipes form an inverted siphon. In the accompanying text, Taccola advises that the ascending pipe in the fountain should be narrower and somewhat shorter than the descending pipe so that the water jet will spring forth to some height. Mariano Taccola, *De Ingeneis,* Palat 766, bk. 3, fol. 28r. Reproduced in Frank D. Prager and Gustina Scaglia, *Mariano Taccola and His Book De Ingeneis* (Cambridge: MIT Press, 1972), fig. 23.

served. It is possible that surveying techniques borrowed from the Islamic world were used in medieval Christendom. Mariano Taccola's drawings show an astrolabe being used for taking levels.[59]

The Bristol agreement specified that Hugh White was to lay his pipes starting at the conduit-heads.[60] The practice of digging pipe trenches by starting at the source may have been common; the Waltham workmen also started at the conduit-head and dug their trench in the direction of the abbey, laying the pipes (and closing the trench?) as they went. Work at Waltham was suspended during the first winter and could be temporarily

FIG. 3.7. Surveyor taking levels with an astrolabe while surveying a watercourse. Mariano Taccola, *De Ingeneis,* Palat 766, bk. 4, fol. 58r. Reproduced in Frank D. Prager and Gustina Scaglia, *Mariano Taccola and His Book De Ingeneis* (Cambridge: MIT Press, 1972), fig. 75.

halted when it rained—sensible precautions, since a waterlogged trench would not provide a stable bedding for the pipe. Working down the slope allowed them to periodically test their pipe as they went along (although they must have had a way of shutting off the flow of water while they were working).

The most common forms of medieval pipe trenches were simple earth-dug trenches and stone-lined culverts. The cross sections of the unlined trenches vary. At Lichfield the pipe was laid in a U-shaped trench, whereas at Kirkstall Abbey the trenches were deep and V-shaped. Lined culverts are known from many sites and were undoubtedly built to provide additional protection for the pipe. At Bordesley Abbey a lead pipe may have been protected by a wooden channel. Most commonly, channels were lined with masonry and roofed with stone slabs, since removable capstones left the pipe easily accessible for inspection and repairs.[61]

At Exeter the fourteenth- and fifteenth-century pipes were laid in subterranean stone-built passageways and rock-cut tunnels. The passages (which were large enough to walk through and are still partly accessible) may have been designed to enable workmen to repair the pipes without disrupting traffic in the busy streets overhead. The "Underground Passages" are thought to represent the Cathedral Close conduit of 1347–49 and the civic conduit of 1420. The construction crew for the former included laborers (who did the digging), sawyers, carpenters (making centerings), plumbers (who made and probably laid the pipes), and masons (paid for their work of covering over the pipes with stones). The materials used in the construction of the passages included stones from various quarries, roofing stone, lime, nails (for the centerings), tallow, wax, pitch, rosin, tin, solder, lead, clay, sand, firewood (for the pipes), broom, rope, and an oak timber beam.[62]

Within the trench or culvert, pipes were frequently provided with some sort of bedding. This helped protect the pipes against fractures by supporting them evenly throughout their lengths and reducing the chances of any movement. Curved roof tiles were apparently used as the seatings for lead pipes at Bordesley Abbey. Some of the earthenware pipes found at Polsloe Priory in Exeter had adhering slate fragments. Whether in unlined trenches or masonry culverts, medieval pipes were often bedded or jacketed in puddled clay, and occasionally mortar, both Roman techniques. It is difficult to say how well the practice worked—some clays can exacerbate pipe corrosion, and any shrinkage of the clay might cause the pipe to fracture. At Saint Augustine's, Canterbury, the fill immediately around the earthenware pipe was sticky brick earth—only the upper trench fills contained rubble. This suggests that the immediate fill around the pipe may have been deliberately selected (as in modern practice) to reduce the chance of damage to the pipe. The top of the pipe might also receive extra protection: at Kirkstall Abbey a triangular arrangement of pitched sidestones with capstones at the apex covered the pipes.[63]

Special arrangements were made when pipelines had to be taken across rivers, ditches, or other topographical obstacles. The London Charterhouse pipe was strengthened by laying it in a "piece of oak covered with a crest of oak" as it crossed a ditch and was "closed in hard stone" where it passed under the highway. A pipe might be taken across an existing bridge. An early example of this practice comes from Verona, where in 873 Louis II granted the bishop the right to take a water pipe over the public bridge. Similarly, the Worcester Cathedral priory pipe crossed the Severn by means of the bridge. Bridges were also used to carry pipes across moats, as depicted in the Canterbury waterworks plan.[64]

The alternative to a bridge was to carry the pipe under the obstacle, on the principle of the inverted siphon. The pipe-run would descend to the bottom of the obstacle, run across, and ascend back up the other side, like a giant U. Since the system was closed, the pressure of the water would cause the water to rise up the ascending pipe until it reached its original level. The Waltham narrative describes the extra protection provided for the pipe as the trench was carried under an intervening river. The pipe seems to have been made longer than usual, probably so that it would not be necessary to have any joints under the water. It also seems to have been especially strong; it was reinforced using clay and wood to protect it against damage from passing boatmen. The Oxford religious houses apparently piped their supplies of water under the various branches of the Thames; the Lacock Abbey pipes were probably taken under the Avon and the Reading pipes under the Kennett. The Exeter system crossed the Longbrook twice, whereas the London Greyfriars' pipe passed under the water by Holborn Bridge. Inverted siphons could also be used to traverse low-lying ground: at Chester the Dominican pipeline must have employed an inverted siphon, as the route crosses land about five meters below the level of the friary. Large-scale inverted siphons, although apparently rare in the Middle Ages, were not entirely unknown; they were used to carry pipes across valleys in order to reach hilltop towns like Siena and Orvieto.[65]

INTERMEDIATE OUTLETS

Medieval pipes were often provided with various intermediate outlets between the conduit-head and the terminus. The study of these structures is complicated by the terminology of surviving records: some words may have been used for more than one type of hydraulic component, and some components may have had more than one name. Unless medieval texts give descriptive details (or can be interpreted in the light of archaeological evidence), it is not always possible to identify the subsidiary hydraulic components they record. Intermediate components on pipe runs are probably underrepresented in archaeological publications, since most excavations tend to concentrate on the buildings near the ends of pipe systems or (more rarely) the conduit-heads, rather than on the main pipe runs.

The Waltham narrative describes the construction of sixteen features located at intervals along the length of the pipe. Each feature is numbered (in a single sequence, starting at the end closest to the conduit-head) and is called either an *expurgatorium* or a *suspirale,* terms that are apparently used interchangeably. The structures were made by the same construction crew as the rest of the conduit and seem to have been built sequentially as the work progressed. The London Charterhouse plan shows a series of what appear to be small square tanks at intervals along the pipelines. The annotations label the tanks *suspirels,* though one is called a *spurgell.* The description of the London Greyfriars' water system mentions three *spurgella,* which were subterranean structures. At least two of the spurgells seem to correspond to small squares shown on a seventeenth-century plan of the water system, which had passed to Christ's Hospital following the suppression of the friary. The squares are probably the *cesperills* referred to in Christ's Hospital records. Thus, *expurgatorium* and *spurgellum* seem to be terms equivalent (at least approximately) to *suspiral.* All three words seem to have been used for outlets (probably in the form of small tanks) spaced at fairly regular intervals along the supply pipe, although *suspiral* may also have been used for vent pipes.[66]

Whatever they were called, intermediate outlets along pipelines could serve a variety of functions (although which were uppermost in the minds of their builders is not easy to determine). The great care taken in describing the exact locations of the Waltham, London Greyfriars, and Charterhouse outlets suggest that it was essential for future workmen to be able to relocate them and that they were not necessarily immediately visible. That some buried outlets seem to have been roofed with removable slabs or planks confirms that ease of access was important. One explanation is that they functioned as inspection chambers: workmen could isolate problems in the system by checking the "manholes." The outlets may also have served as shutoff points when repairs to the pipes were necessary.[67]

A second function of intermediate tanks seems to have been water purification. Like the tanks at the conduit-heads, many intermediate cisterns apparently served as settling tanks, and some were equipped with

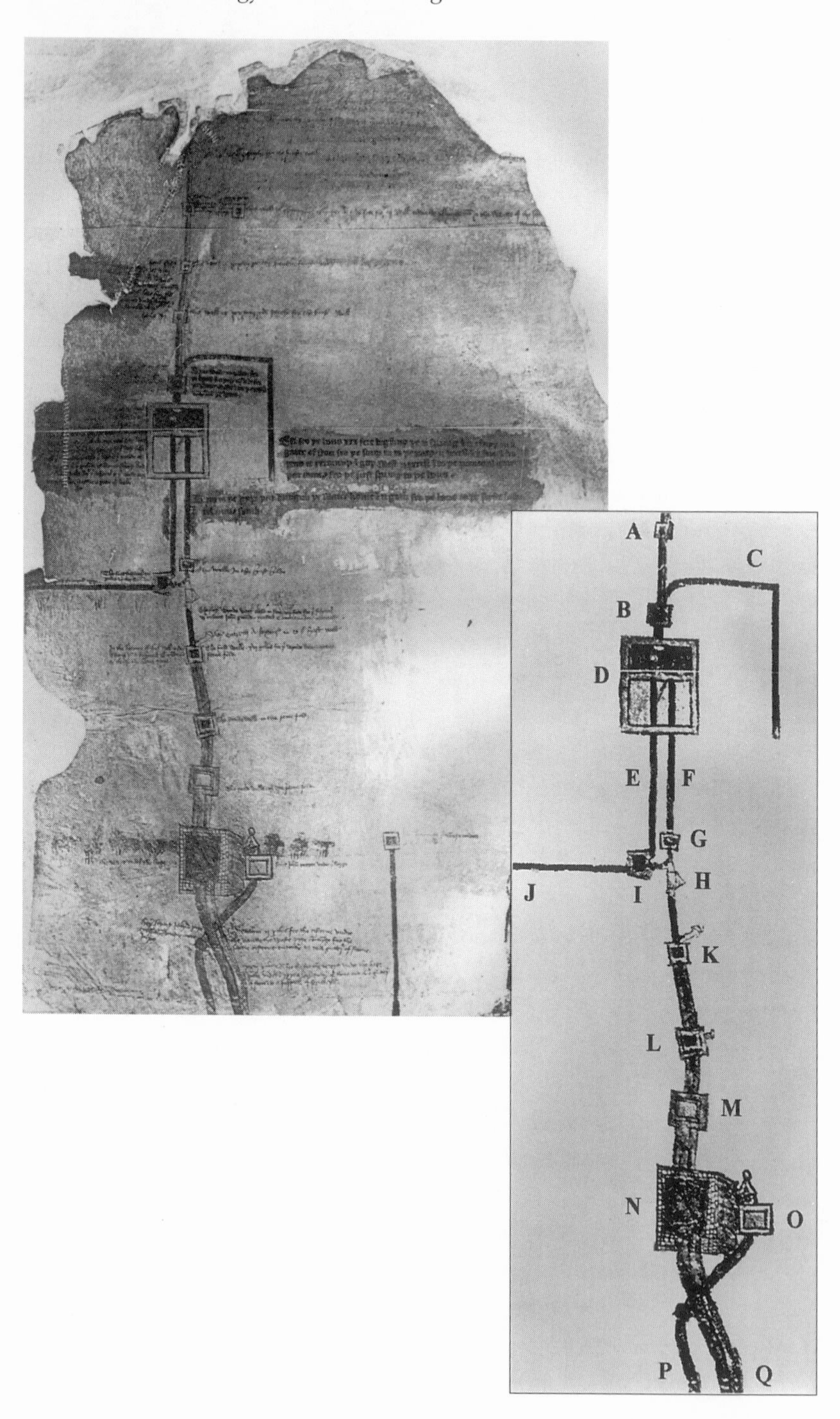
A
B
C
D
E
F
G
H
I
J
K
L
M
N
O
P
Q

perforated filters to screen out suspended debris. A series of five (apparently) open rectangular tanks, spaced at intervals along the main pipe run between the conduit-head and the Cathedral precinct, are shown on the Christ Church, Canterbury, plan and diagram. The tanks are rectangular and are connected by a sequence of offset intake and outtake pipes. In addition, each is equipped with what appears to be a short length of pipe with a round terminus. The drawings are thought to represent settling tanks equipped with purge pipes. A well house that initially served as a distribution point for the Wolvesey Palace water system in Winchester seems to have been adapted to serve as a settling tank when the system was extended. The London Charterhouse plan describes a "susprall with a tampion to cleanse the home pipe."[68]

FIG. 3.8. London Charterhouse plan, first membrane (detail). This section of the plan shows part of the complex head of the water system, with two large rectangular conduit houses and a series of small intermediate features. The notes on the plan identify these structures as wells (*A, B, G, K, L, M*) and/or suspirels (*I, K*). Feature *H* is described as a wind vent. Channel *C* conveyed springwater to well *B*. From the lead cistern in the north half of conduit house *D*, two pipes carried the water south. Pipe *E* is identified as a waste pipe, which emptied waste water into suspiral *I* and brick gutter *J*. Pipe *F*, the Charterhouse "home pipe," conveyed the water to conduit house *N*. Between wells *K* and *M* the pipe is double; between *M* and *N* the plan shows a triple pipe. Conduit house *N* has a cistern that completely fills the brick building and is flanked by a small "receyte" that belonged to St. John's Hospital. The St. John's pipe (*P*) and the Charterhouse Home Pipe, together with a parallel small waste pipe (*Q*) cross each other just south of conduit house *N*. A bit to the east of features *N* and *O* lay the head of the conduit, which supplied the nuns of Clerkenwell (not shown). Muniments of the governors of Sutton's Hospital in Charterhouse, London, MP 1/13. Mid-fifteenth century. Reproduced in R. A. Skelton and P. D. A. Harvey, *Local Maps and Plans from Medieval England* (Oxford: Clarendon Press, 1986), pl. 19.

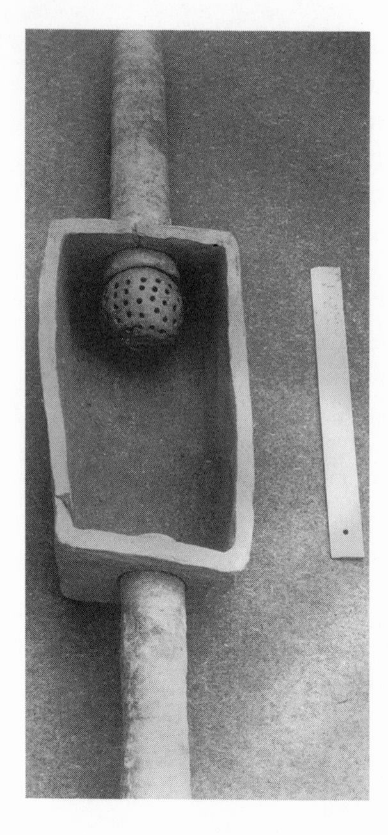

FIG. 3.9. Stone inspection chamber with earthenware pipes. The outflow pipe is fitted with a perforated filter. Kloster St. Johann, Alzey. Frontinus-Gesellschaft, *Die Wasserversorgung im Mittelalter,* Geschichte der Wasserversorgung, no. 4 (Mainz am Rhein: P. Von Zabern, 1991), fig. 25, p. 35. Photograph by Ch. v. Kaphengst.

In one of the manholes at Wells, the pipe was fitted with a washout valve, so that any sediments in the pipeline could be flushed out. The Canterbury waterworks plan shows purge pipes (labeled *purgatorium* or *purgatorium calami*) in several places, especially in the elbows formed when feed pipes bend upward to supply the water towers. These were apparently fitted with stopcocks and could be opened to flush out the main pipelines. Cleaning sediments out of pipes was also undertaken by means of wires: in the mid-fourteenth century five fathoms of wire ("teys de wyr") were purchased to clean out the pipe of the conduit in the king's mews at Westminster.[69]

Reduction of pressure seems to have been another function of the intermediate tanks. A series of open tanks can reduce the head of water, and therefore the static pressure, in a piped system. Medieval engineers seem to have had an empirical understanding that the provision of intermediate tanks would reduce pressure in the pipes, although they may have thought that the danger came from compressed air that needed to be released by means of a "breathing hole."[70]

FIG. 3.10. Diagram of the water system at Christ Church, Canterbury (c. 1153–61). The head of the system is in the top left corner. Five rectangular settling tanks, which appear to be equipped with purge pipes, are spaced along the pipeline as it passes through fields, a vineyard, and an orchard. Trinity College, Cambridge, Ms. R.17.1, fol. 286r. Reproduced in R. A. Skelton and P. D. A. Harvey, *Local Maps and Plans from Medieval England* (Oxford: Clarendon Press, 1986), pl. 1B.

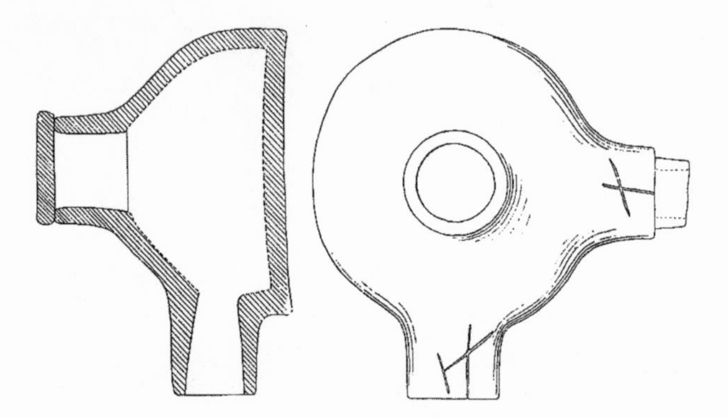

FIG. 3.11. Ceramic junction box, Glenluce Abbey. Tally marks were scored into the surfaces of the junction boxes and the earthenware pipes. Junction boxes were employed when the pipes changed direction; the removable lids permitted them to serve as inspection chambers. Stewart Cruden, "Glenluce Abbey: Finds Recovered during Excavations," *Transactions, Dumfriesshire and Galloway Natural History and Antiquarian Society* 29 (1950–51): fig. 21.

Besides relieving static pressure, intermediate tanks could generally reduce stress at vulnerable points in the system by isolating disturbances in the flow and slowing down the movement of the water. Small tanks served as junction boxes when the pipeline changed direction or when additional intake or outtake pipes were joined to the system. At Kirkstall Abbey small stone-lined tanks were situated at the points where the pipe changed direction. The Wolvesey Palace well house also served as a junction box at a corner in the pipeline. The earthenware pipes at Glenluce Abbey were fitted with small ceramic junction boxes at the points where the line changed directions. These circular junctions would have protected the pipes from the current's inertial thrust as the water rounded the bend. (The junction boxes had tally marks similar to those in the pipes and had removable lids. The lids meant that they could also be used as inspection chambers, and sediments could be cleared out through the lid.) Not all pipe junctions took place in junction boxes, however: the Canterbury plan shows pipes with direct T- and Y-junctions, without any indication of a box or tank.[71]

Finally, some intermediate tanks seen to have served as dipping places, access points where water could be drawn. The Wolvesey Palace well house was fitted with steps to provide easy access to the monolithic trough containing the water. The suspirals in Coventry's municipal conduit were apparently being used by private citizens to draw off water, to the detriment of the public supply: city officials variously attempted to regulate, tax, and block the structures.[72]

OPEN CHANNELS

The alternative to using pipes for conveyance was to employ some kind of channel. Channels were built by a variety of craftsmen, depending in large measure on the type of lining. Laborers generally did the digging, carpenters were responsible for timber-lined channels and planked covers, and masons built stone-lined channels and vaults. As in the case of pipes, the main factor seems to have been the workmen's experience in handling the specific materials: hydraulic expertise was not the monopoly of any one craft.[73]

Medieval channels seem generally to have been used when the quantity of water was a more important consideration than its quality. They were frequently supplied by river or stream diversions. Drainage channels received additional water in the form of storm runoff and the overflow from intake systems. The water flowing through the drains and sewers carried away domestic and industrial wastes. When a fluvial source was unavailable, rainwater might be collected to flush out the drains. Some drains were fed by water from moats or fishponds or even flushed by the tide.[74]

Most channels did not run full. The depth of flow in a channel could vary considerably; drains and sewers in particular were likely to have diurnal and seasonal variations in flow, depending on patterns of use and fluctuations in storm-water runoff. The uncertainty in reconstructing past levels of water complicates the hydraulic analysis of historical channels. In order to calculate the approximate discharge, it is necessary to estimate the normal depth of flow. The basic principles of open-channel hydraulics can be expressed in flow equations, such as the widely used Manning formula, although more complex flow problems may require empirical studies. The quantity of water delivered by a channel is a product of the cross-sectional area and the mean velocity of the current. The velocity of the current is, in turn, determined by the slope and by the retarding effect of surface friction with the channel walls. Thus, the factors that affect the efficiency of channel flow are the gradient, the roughness of the channel walls, and the dimensions and shape of the cross section. Since the necessary data can be retrieved archaeologically (within acceptable margins of error), the hydraulic analysis of surviving medieval channels is an attainable (if as yet virtually unrealized) goal.[75]

The maximum average gradient for a channel (as for a pipe) was determined by the difference in elevation between the source and the terminus. Gradients could be reduced by following a sinuous course rather than a direct line. Within a single channel there might be changes in gradient; the depth and velocity of current varied with the profile, running fast and shallow down steep slopes, deeper and more slowly when the slope was gentle. When channel gradients change suddenly, flow disturbances occur: the localized turbulence of a hydraulic jump can result in extra damage to the channel walls. The designers of channels had to make a choice: channels with steep gradients and high velocities were subject to greater wear and could require more frequent repairs; channels with shallow gradients and low velocities were subject to greater sedimentation and could require more frequent cleaning.

Systematic studies of the longitudinal profiles of medieval channels are not available. Regrettably, archaeological publications seldom report channel gradients (and often omit even the direction of flow). Recent studies of Roman aqueducts have analyzed the variations in their profiles (an aspect of channel engineering obscured by the traditional practice of publishing only overall average gradients). A similar approach to the study of medieval channels would be highly desirable. The medieval channel gradients that have been published reveal considerable variety: 0.8 percent at Saint-Denis, 2.7 percent at Chester, and a whopping 38 percent (21 degrees) at Worcester Priory. In the cases of some monastic drains, steep gradients may have been deliberate attempts to increase the flow velocity to self-scouring levels in order to reduce the need for cleaning. In the Sienese *bottino* for Fonte Gaia, deliberate steps were taken to reduce the gradient of the intake channel. By cutting a sinuous channel in the tunnel floor, a steady slope could be maintained, and the increase in the length of the winding channel relative to the fall slowed the flow of water. Following in the Roman tradition, aqueduct bridges and tunnels were sometimes used to maintain channel gradients, carrying them across or through intervening topographical irregularities. As they crossed low-lying areas, masonry channels could be carried on top of arcades, as at Salerno and Sulmona.[76]

The walls of a channel offer frictional resistance to the flow. The amount of resistance depends on the channel's hydraulic radius (the ratio between the wetted perimeter and the cross-sectional area of the portion of the channel actually filled by water) and the roughness of its walls. The most efficient cross section for open-channel flow is a semicircle, or the lower part of a U-shaped channel. Trapezoids are also efficient; square or rectangular channel cross sections are relatively inefficient. If a channel is rectangular, the optimum flow is obtained when the depth of the water is equal to half its width.

The great majority of medieval masonry channels had rectangular or square cross sections, probably because this was the easiest type to build. In some cases, however, channel cross sections seem to have been deliberately designed for greater hydraulic efficiency. At Chester's Dominican Friary, the cross section of the main drain changed from square to trapezoidal part-way along its course. The square portion is built of simple stone blocks and flat floor slabs; the trapezoidal section required special chamfered blocks and hollowed-out floor slabs. The fills in the two different sections are revealing: in the square drain only a fine layer of silt was found, whereas in the tapered portion the bottom was filled with sewage. It appears that the change in shape was a deliberate attempt to improve the efficiency of the flow in precisely that portion of the drain that was associated with a latrine and hence was more vulnerable to blockages. At Chelmsford Dominican Priory, an intake culvert was rectangular, but the reredorter drain had walls that tapered in toward the bottom.[77]

The late-twelfth-century rectangular drain at Norton Priory was replaced by a drain with a semicircular channel and a steeper slope in the fourteenth century, in the segment that passed beyond the latrine block. The very steep Worcester reredorter drain also had a semicircular channel. At Norton Priory and Worcester, both the gradients and the channel shapes may have been designed to compensate for a low volume of water. Increasing the velocity of the flow would improve the self-scouring capacities of the drains, particularly in the sections that were required to flush away sewage.[78]

The degree of frictional resistance to the flow in a channel depends not only on the hydraulic radius but also on the roughness of the internal

FIG. 3.12. Salerno, medieval aqueduct arcades. The arches carried a masonry water channel on top and were employed to maintain steady gradients across topographical irregularities. Armando Schiavo, *Acquedotti romane e medioevali* (Naples: F. Giannini, 1935), fig. 9, p. 45.

surfaces. The roughness of the materials used to line medieval channels varied considerably. Many channels were simply unlined, open ditches. A grid of open drainage ditches formed the first water-management system at Norton Priory and was associated with the temporary timber buildings of the twelfth century. Linings were used to reinforce the sides of a channel and in some instances to keep the flow of water from undermining nearby buildings. Linings also helped protect the purity of the water supply in intake channels and provided an easier surface to clean in the case of drains. At the Austin Friary in Leicester, the main drain consisted of an open ditch screened by a wattle fence until the fifteenth century, when it was finally replaced by a masonry drain. The use of inexpensive wattle or timber revetments for drains and channels was probably quite extensive, although the chances of preservation for such organic materials are relatively poor.[79]

Better preserved are masonry channels. These were usually made of inexpensive local materials, such as sandstone, gritstone, limestone, chalk, brick, even roof tiles, though on occasion nonlocal materials such as Caen stones and Purbeck marble were used. When expensive imported materials were employed, they probably were either *spolia* (reused building materials) or leftovers from other building projects, except when they were used to build arched openings in building foundations. The material used for the channel floor might differ from the walls—some masonry channels have unlined gravel or clay floors, and many have stone paving slabs. The degree of care taken to ensure a smooth internal surface varied widely: some channels were faced with roughly dressed stone, others with smooth ashlar. A garderobe drain at Southampton Castle had carefully smoothed ashlar in the bottom courses, but the upper masonry was left rough. It appears that the builder was empirically aware that smooth walls would improve the flow (and the self-scouring capacity) of the drain, but he reverted to a cheaper option once safely above the water level.[80]

Channel stonework was usually bonded with mortar but in some instances was packed with clay. On most sites the mortar was apparently ordinary lime mortar, although occasionally more specialized hydraulic mortars were employed. According to documentary sources, medieval cement recipes for cisterns and wells sometimes included crushed tile, fat, or egg. Some excavated water channels are known to have had crushed tile or brick added to the mortar, creating a cement similar to Roman *opus sig-*

ninum. A molten "cement" made of one or more of ingredients such as pitch, wax, resin, crushed tile, egg, and sulfur was used for cisterns or when the masonry was particularly exposed to wet conditions. There seems to be little archaeological evidence for the application of multiple layers of polished mortar in the Roman fashion, but little direct work has been done on the subject. Clay linings have been reported on a few sites and may have been attempts to create a smoother internal surface or to alter the dimensions or cross-sectional profile of the channel. A gritstone channel at Fountains Abbey was lined with timber and sheets of lead.[81]

Many channels were covered. In the case of intake channels, roofs helped protect the purity of the water; in the case of drains, the covers helped contain noxious odors. Documentary sources refer to the use of wooden planks for this purpose, although few have survived. The master and brethren of Saint Bartholomew's Hospital, London, received a royal license in 1297 to cover, "on account of its excessive stink," a watercourse that ran through the middle of their hospital. The cover was to be made of wood and stone. In 1259 a new covered channel was constructed to convey the waste from the Westminster Palace kitchens to the Thames, "which the king ordered to be made because of the stench of the dirty water which was carried away through his halls, which used to infect persons spending their time there." The building accounts for work on this "great gutter" include payments to the carpenters who covered it with planks. A very common method of roofing channels was to use stone slabs as capstones. The technique was an old one, known from ancient and early medieval sites as well as many later medieval contexts. Local availability of a suitable material was probably the main determinant of the type of stone used; the capstones ranged from large, roughly hewn blocks to thin flags. Small channels might be covered with roof tiles. Some channels were vaulted, and many flowed through arches when they passed under building foundations. Since channels did not flow full, the type of cover did not affect the hydraulic efficiency.[82]

When changes in a channel's dimensions or direction occurred, the transition was frequently an abrupt (often a right) angle. Branch channels often joined a main channel with a simple T-junction (though often entering at a higher level than the bed of the main channel). Such abrupt transitions were similar to Roman practices, but on some sites attempts were made to ease the current into its new state of flow. At Cuckoo Lane,

Southampton, one branch drain junction forms a right angle with the main stone drain, but another junction is more oblique. Oblique branch-drain junctions also occur at Furness Abbey. The transition to a narrower channel in the Southampton Castle garderobe drain is eased by inserting angular blocks in the corners of the walls at the point where the flow is constricted.[83]

Building costs seem to have been an important element in the design of most medieval channels. Many channels were left unlined or were lined with wattle or wood. Even masonry channels were often built of inexpensive local materials, with relatively rough internal faces and square or rectangular cross sections. Medieval builders were, nevertheless, capable of making hydraulically efficient channels, with carefully smoothed internal faces and trapezoidal or rounded cross sections. The relative scarcity of such sophisticated channels probably reflects their higher construction costs, but when problematic local conditions required it, technologically advanced solutions were available. Otherwise, most patrons seem to have been content to make do with less costly channels. They might require more frequent cleaning, but in general they worked well enough.

SINTER

The configuration of a channel (or a pipe) could be altered by the buildup of sinter, a calcium carbonate incrustation that forms as a precipitate when the water is hard. Sinter has been studied in Roman aqueducts and pipes but is seldom mentioned in connection with medieval hydraulic installations. It was a widespread problem in the Roman period, and because medieval water systems were built in the same general geologic zones, it is probable that the apparent discrepancy merely reflects the differences in the current state of scholarship on ancient and medieval water systems. Sinter reduces the internal dimensions of pipes and channels, diminishing their flow capacities. If not periodically removed, it can finally clog a system. (Sinter does, however, create a buffer between the water and the lead of lead pipes and cisterns, which reduces the risk of lead leaching into the water.) The neglected Sienese *bottini* now have heavy calcium carbonate incrustations, particularly in places where water pours in through the vertical shafts. One aspect of their maintenance during the Middle Ages was the periodic removal of the incrustation with axes and special hooked instruments.[84]

SEDIMENTATION AND CLEANING

Besides mineral incrustations, channels and pipes could become clogged with sediments. The deposition depended on the rate of flow: if velocities were too low, the system would not be self-scouring and would need periodic cleaning. An analysis of the sediments found in pipes and channels can be a rough guide to the hydraulic efficiency of a system. The frequency with which primary sediments are found in medieval channels indicates that many did not maintain sufficient flow velocities to flush the system, even during peak discharge flows. The sediments recovered archaeologically range from thin lenses of clean silt to thick deposits rich in finds and organic material. The detailed analysis of channel sediments has been undertaken on a number of sites, because such deposits are rich sources of environmental evidence. Close analysis of the sediments can also help in the interpretation of hydraulic structures. For example, samples from a stone-lined drain at Bermondsey Priory in Southwark were taken in three places. Upstream, the samples showed negative results, but within the area of a suspected latrine, as well as downstream from the building, the samples contained quantities of maw worm eggs, indicating that human sewage was entering the system from within the building.[85]

In order to facilitate inspection and cleaning, channels needed to be accessible. Obviously, this was not a problem for open leats and gutters. When channels were covered, several solutions were adopted. In the case of monastic main drains, the channels were often big enough for a workman to walk through (and were provided with some sort of door). As in seepage tunnels, the space required by laborers, rather than the volume of water, seems to have determined the overall dimensions of the channel. These large drains are probably the source of many "secret passage" legends. The Exeter Cathedral fabric rolls of 1340–41 include a payment of 2½ pence for a lock for the door of the drain. This may have been a measure against the use of the drain as a passageway by mischievous boys or other unauthorized persons—the Canterbury choirboys have reportedly haunted the medieval drains within living memory. Some drains had openings in one of the walls, so that the sediments could be removed. Ease of access may account for the widespread popularity of planks and stone slabs, which could be readily removed, as covers for channels. Alternatively, specially constructed manholes could provide access points for inspection and cleaning. At Hartlepool, a late-thirteenth-century stone-lined drain had its cover slabs sealed with clay but was accessible by means of a 20 x 15 centimeter manhole, capped by a moveable cover stone. The use of specially constructed manholes for channels seems to have been relatively uncommon in the medieval period, however, apart from the vertical shafts associated with seepage tunnels. This contrasts with both Roman practice and the abundant evidence for pipeline suspirals.[86]

SLUICES

The flow of water in open channels could be regulated by sluice gates, which were familiar components of fishponds and mill leats and which were similar to their Roman predecessors. Vertical slots that guided the wooden gate are preserved in channel walls on several sites. Exceptionally, the wooden gate or framework may also be preserved. On monastic sites, sluice gates were often situated at the upstream end of reredorter drains and were designed for underflow operation. When the gate was lowered, the level of water in the channel built up behind it; when raised, the sudden surge of water would scour the drain bed. The use of a sluice gate was particularly advantageous if the flow of water was low. Even drains flushed by river diversions could suffer seasonal shortfalls. At Hirsau, where it was the almoner's duty to periodically divert water through the latrine drain, a sluice was used to build up a better head of water in the summer. Sometimes the water for flushing the drains was stored in a tank or cistern before being released. Sluice gates could also control diversions and were used to regulate the flow of water in branch channels. Because of the hydrostatic load, considerable force could be required to raise a gate. Little direct evidence for lifting mechanisms survives, but it is thought that windlasses were probably employed. Surviving records indicate that fishpond sluices were made by carpenters; channel sluices, with their masonry slots and wooden fittings, may have been joint products of carpenters and masons.[87]

EFFLUENT DISPOSAL

The wastes carried by medieval drains were disposed of in a variety of ways. Some drains discharged into pits or soakaways, such as the soakaway associated with a kitchen drain at Guildford Priory. Monastic sewers occasionally fed into fishponds: this seemingly unhygienic practice may actually

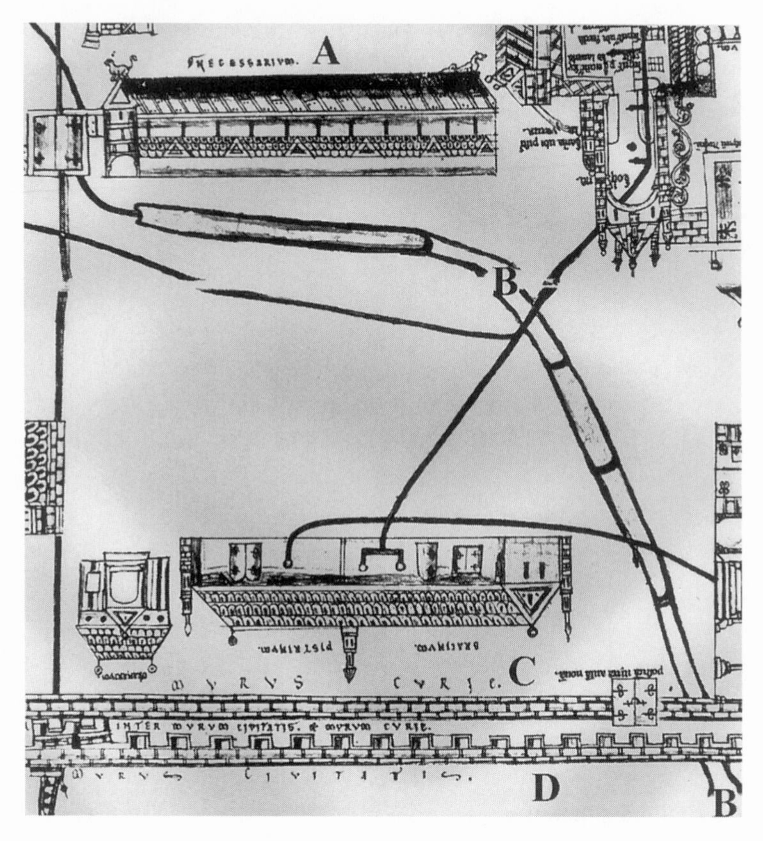

FIG. 3.13. Detail of the Canterbury plan, showing the latrine-block (*Necessarium, A*) and the great drain passing below it (shown as a double line, *B*). A workman cleaned out the accumulated muck every Monday. The drain passed through the walls of the monastery and the town (*C, D*), emptying its sewage into the city moat. The citizens liked to block up the exit during quarrels with the monks. Reproduced in R. A. Skelton and P. D. A. Harvey, *Local Maps and Plans from Medieval England* (Oxford: Clarendon Press, 1986), pl. 1A.

have been beneficial, inasmuch as excreta can increase yields by nourishing the plankton on which the fish feed. Some drains emptied into millponds: the millwheels would have provided (inadvertent) wastewater treatment as they aerated the sewage. The Norton Priory reredorter drain fed into the millpond; in Dublin one medieval mill was known as Muileann a Chacca (Shitty Mill) and another bore the name Schytclapp Mill. Drains also discharged into streets or the sea, but most commonly they emptied their effluents into a convenient moat, ditch, or river. The main drain of Christ Church, Canterbury, discharged its effluent into the city moat. When the quarrels between the townsmen and monks became especially heated, the citizens would obstruct the end of the drain. In Carlisle the Dominicans' drain emptied its filth onto the land of Carlisle Priory just outside the city wall. The ensuing quarrel simmered for much of the thirteenth century, and the canons took various forms of direct action when other means of persuasion failed. They tried erecting withies to control the outflow and finally piled up a five-foot-high mass of beams and stones to block the channel. This had the gratifying result of backing up the sewage into the friary and rendering it "offensive, unhealthy, and uninhabitable," but it also threatened the safety of the town: the obstruction made it easier for the Scots to scale the city wall. Not only could drains lead to such unedifying quarrels, but the pollution of the watercourses into which they debouched also made it all the more desirable to have separate, spring-fed intake systems.[88]

DISTRIBUTION

Upon reaching its destination, water was made available to users at some sort of delivery point. In nonpressure systems, the water simply flowed into some kind of receptacle. In pressure systems, rising pipes supplied water towers, some types of fountains, and standpipes on the principle of the inverted siphon. Simple water systems might discharge directly into a single basin; others had more complex distribution networks and structures.

There were several types of delivery structures. Most, such as tanks, standpipes, and various kinds of fountains, were designed to supply users with water that they could carry away in containers and use elsewhere. Some fountains (especially monastic lavers) served for washing hands and faces. Occasionally water was fed directly into special-use structures, such as kitchen sinks, laundry basins, watering troughs for animals, industrial vats, and even bathtubs. Dipping tanks and fountains could themselves double as distribution points, feeding additional branch lines with their excess water.

Roman aqueducts typically discharged into a distribution tank known as a *castellum divisorium,* located at the edge of a city. Here the water was divided into the separate branch pipes that formed the urban distribution network: the elevation of the *castellum divisorium* established the head of

FIG. 3.14. The water tower, Christ Church, Canterbury. The intake pipe rose up through a central column to fill the upper-story fountain. The fountain served as a *lavatorium,* and its outlet pipe fed the next fountain by means of an inverted siphon. Robert Willis, "The Architectural History of the Conventual Buildings of the Monastery of Christ Church in Canterbury," *Archaeologia Cantiana* 7 (1868): fig. 7.

water for the pressure system. The branch pipes from the initial *castellum* might feed secondary elevated *castella,* or water towers, which maintained the head while serving as junction boxes for further branching.[89]

Known medieval water towers show little architectural affinity to Roman *castella,* but the hydraulic principles are similar. The London Charterhouse plan shows a polygonal two-storied conduit house (which the plan calls an *age*) in the center of the cloister garth. According to the annotations on the plan, "the main pipe comes and rises up into the age in the midst of a fair square cistern of lead." The Charterhouse *age* served as a water tower and a distribution reservoir: the plan shows four branch pipes issuing out of its base. The *age* does not seem to have been designed to provide immediate access to the water; a door at ground level opened to reveal the pipes, but the upper-story cistern could be reached only by means of a ladder. Similar central distribution structures are thought to have stood at Mount Grace and Witham, both Carthusian houses.[90]

At Canterbury the twelfth-century water tower depicted on the Christ Church plan still survives in part. The pipe rose through a central pillar into a second-story fountain, which was used as a laver but also served as the reservoir that established the initial head of water for the distribution network. In contrast to the Charterhouse *age,* it did not function as a distribution reservoir. A single outlet pipe descended from the basin to feed the next fountain by means of an inverted siphon.[91]

DIPPING PLACES

A simple form of delivery structure was provided by sunken cisterns or dipping places. At Kirkstall Abbey sunken masonry cisterns provided points where water could be drawn in the cloister and in the Warming House courtyard. The rectangular cisterns were provided with inlet and outlet conduits (the Warming House cistern was fed by a pipe, though this seems to be a secondary feature), and each had steps on one side to facilitate access. At Polsloe Priory a sunken rectangular cistern was supplied by an earthenware pipe; access was provided by a flight of descending stairs. The well house at Wolvesey Palace, Winchester, was a pipe-fed sunken cistern of the same general type and was also entered by descending a short flight of steps. A subterranean room in the cloister garth of Wells Cathedral is known as the Dipping Place. It is covered by a barrel vault and approached down a flight of steps: the conduit flowed through the center of the chamber, so that water could be easily drawn out. With the possible exception of Polsloe Priory, these dipping places do not seem to have been terminal structures; their outlet conduits apparently conveyed water to other parts of the system.[92]

FOUNTAINS

Medieval fountains generally took the form of spill fountains or splash fountains. Spill fountains simply discharge water through one or more outlets into a receptacle (often a trough). Some medieval examples were architecturally elaborate, set in wall niches or vaulted, and had multiple spouts and troughs. Most Sienese civic fountains were spill fountains, fed through spouts supplied by the *bottini* behind them. The basic fountain

FIG. 3.15. Freestanding splash fountain. The water rises by means of an inverted siphon up a pipe in the central shaft into an enclosed upper basin. From there it flows out through spouts into the basin below. Titus Livius, Mailand, Ambrosiana MS C. 214 inf., fol. 12, 1372–73. Reproduced in Bernhard Degenhart and Annegrit Schmitt, eds., *Corpus der Italienischen Zeichnungen, 1300–1450,* vol. 2, no. 3 (Berlin: Gebr. Mann Verlag, 1980), pl. 38b.

type is reminiscent of Greek fountain houses. Those drawing water would dip their vessel in the trough or hold it below the spout.

The second common type of medieval fountain was the freestanding splash fountain. In its usual form, a vertical feed pipe rises up the center of a column or shaft and discharges its water into a basin (the water rising through the pipe by means of an inverted siphon). Splash fountains may have tiers of basins, with water cascading from basin to basin through spouts or taps. The shapes of the basins, elaborations of the shaft, and sculptural details allowed for great decorative possibilities. As in the case of spill fountains, vessels could either be dipped directly into the open basin or held under a spout.

In cloisters splash fountains were usually enclosed in a small room known as a fountain house and served as washing places (*lavatoria*). The fountain house not only sheltered users from inclement weather but also protected the water in the fountain from pollution and provided some insulation against freezing. Cloister fountain houses with splash fountains

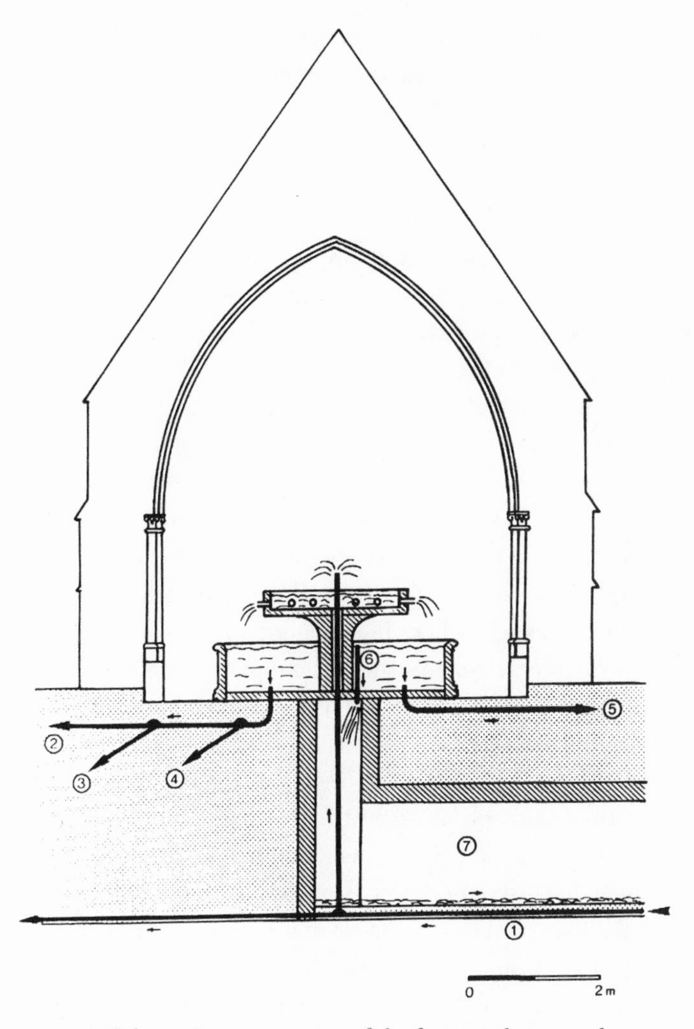

FIG. 3.16. Schematic cross section of the fountain house and *lavatorium* at Maubuisson, which served as the main distribution point for the convent's water system. Hypothetical reconstruction. *1,* intake pipe; *2,* branch pipe to kitchen? *3,* branch pipe to nuns' quarters; *4,* branch pipe to *cellarium; 5,* branch pipe to *mandatum* basin (for ritual foot-washing); *6,* overflow; *7,* vaulted drain. Frontinus-Gesellschaft, *Die Wasserversorgung im Mittelalter,* Geschichte der Wasserversorgung, no. 4 (Mainz am Rhein: P. Von Zabern, 1991), fig. 31, p. 217.

are known from a few English monasteries (and remained popular on the Continent); but from the beginning of the thirteenth century, recessed troughs in wall niches became the most common type of English laver. The reason for the change in fashion is not known, though the new form was less vulnerable to frost damage and was cheaper to build. The troughs were fed by overhead tanks (themselves supplied by feed pipes), which discharged the water through a row of taps or spouts.[93]

English civic fountains (known, confusingly, as conduits or conduit houses) show a similar preference for elevated, pipe-fed cisterns with taps, though unlike wall lavers they were often freestanding. Water was drawn from the tap, not scooped out of an open basin, though the taps may have been kept open so that the water ran continuously, at least by day.[94]

The overflow from fountains could be used to feed secondary basins, troughs, or pipelines. The water was carried away by means of a waste pipe, with the mouth set at a level above the bottom of the basin. The Canterbury plan shows the laver waste pipes as freestanding vertical pipes in the middle of the basins; elsewhere they might be fed through an aperture in the basin wall. In some systems the waste pipe from one fountain became the feed pipe of the next, as the water flowed from fountain to fountain in a series of inverted siphons.[95]

Fountains and lavers were made of various materials. Textual evidence shows that the basins of lavers and fountains could be made of metal, though the survival rate for such metal components is very low. The bronze Marktbrunnen in Goslar is one surviving example. King John had given the canons at Waltham Abbey a tin laver that had formerly been installed in Westminster Palace, and lavers of tin appear in the Dissolution inventory for the Austin Friary at Ludlow. The late-fourteenth-century lampoon "Pierce the Ploughman's Crede" describes the London Dominican cloister's "conduits of clean tin" and "lavers of latten." The "tin" may have been a tinned copper alloy. Latten was brass, or a similar alloy of copper, zinc, lead, and tin. Bethlehem Hospital had a brass laver in the cloister, and lead troughs or basins may also have been used.[96]

Perhaps more typically, fountain basins were made of stone. Purbeck marble fragments of molded panels, a central lobed basin, and a trough are thought to have belonged to a later twelfth-century fountain at Westminster Palace. The civic fountains in Viterbo and Siena were both made out of common local building materials: Viterbo's fountains were made

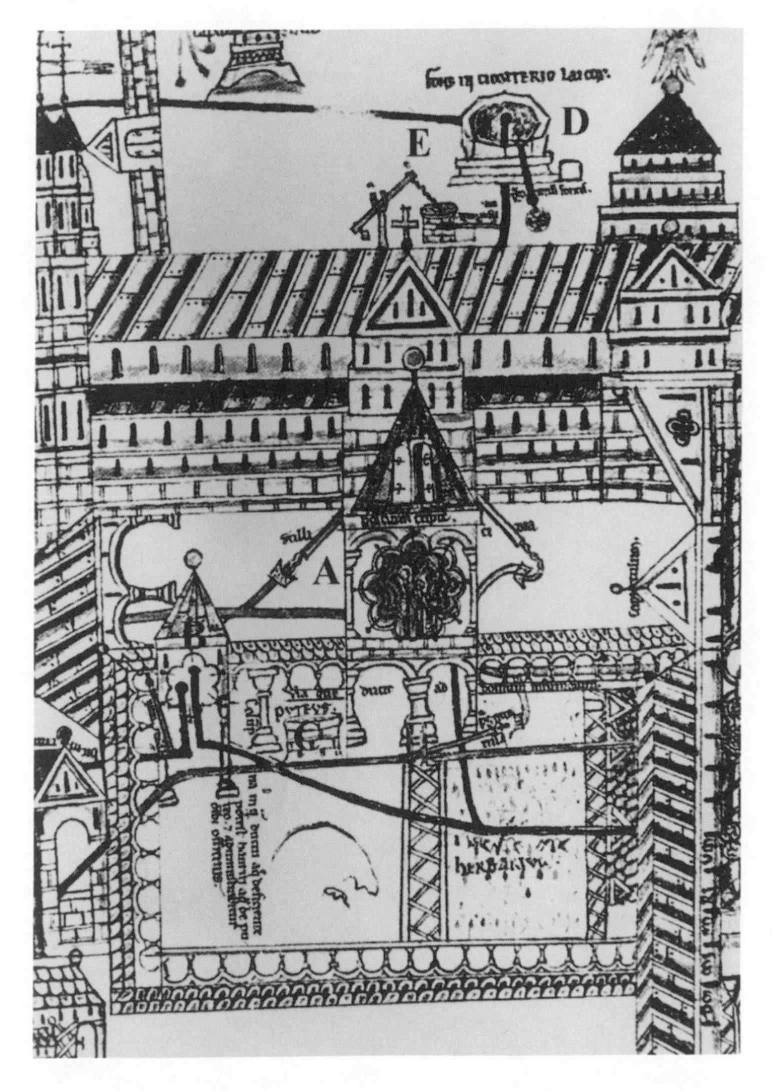

FIG. 3.17. Canterbury plan, detail. The water tower (*A*) is the structure in the center, with a large lobed fountain in its upper story. Below and to the left of the water tower is a smaller lobed fountain (*B*), which served as the infirmary laver. Between the two stand a column shaft and a well (*C*). According to the note on the plan, water drawn from the well and poured into the column could be distributed to all offices when the aqueduct was out of order. At the top, to the far side of the church, stand a polygonal fountain (*D*) in the lay cemetery and a well with a shadoof (*E*). Reproduced in R. A. Skelton and P. D. A. Harvey, *Local Maps and Plans from Medieval England* (Oxford: Clarendon Press, 1986), pl. 1A.

FIG. 3.18. View of Westminster Hall, by Wenceslaus Hollar (1647). The late medieval fountain under the canopy is Westminster Palace's Great Conduit. The construction of such elaborate fountains could require the services of many different kinds of craftsmen. Reproduced in Katherine S. Van Eerde, *Wenceslaus Hollar: Delineator of His Time* (Charlottesville: Univ. Press of Virginia, 1970), fig. 12.

of volcanic *peperino* (tuff), whereas those in Siena were built of brick and stone.[97]

Construction of a fountain could require the services of several specialist artisans. The Durham Abbey accounts for 1432–33 record payments for making a new laver. The expenses include payments to several masons for quarrying marble at Eggleston; for transporting, working, and polishing the stone; and for building a lodge for the stone working. Other craftsmen who were paid for their work on the laver included Thomas the plumber (who probably installed the pipes) and Laurence the latoner (lattener) (for making the spouts). Many different masters were employed in the construction of Perugia's fountains, including Fra Bevignate, who was the overseer of the project; the Venetian magister Boninsegna, who seems to have been the chief hydraulic engineer; and the famous sculptors Nicola and Giovanni Pisano and Arnolfo di Cambio. An artisan capable of making an elaborate laver could command a good return for his services. In 1288 John the Potter (a bronze-founder) agreed to ride down to Ramsey Abbey from Huntingdonshire, along with two of his journeymen, to make a new laver "of good and durable metal," thirty feet long and two and a half feet high. In exchange, the abbot agreed to pay John thirty pounds and a gown, with

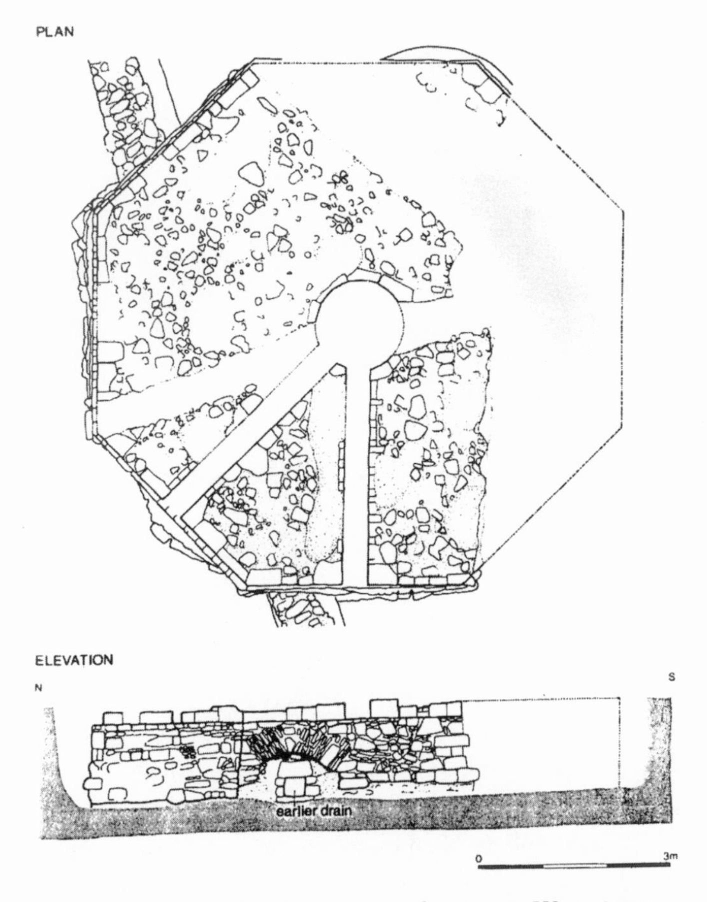

FIG. 3.19. Octagonal foundation for the fountain in Westminster Palace Yard known as "le Standard" and later as the Great Conduit, as revealed in the course of archaeological excavations. The pipe channels are visible in the masonry. Valerie Horsman and Brian Davison, "The New Palace Yard and Its Fountains: Excavations in the Palace of Westminster 1972–4," *Antiquaries Journal* 69, no. 2 (1989): fig. 6, p. 289.

one-third of the sum to be given to him in advance to purchase materials. In addition, the abbot was to provide food for the men and their horses while the job lasted. The men were each to receive a daily ration of two loaves of bread, two gallons of beer, and a dish of meat or fish. John, as the master craftsman, was to have his beer drawn from the convent cask, and one of his loaves was to be "monk's bread." His men's bread and beer, however, were to come out of the servants' hall.[98]

TAPS AND STOPCOCKS

Medieval taps and stopcocks, like their Roman predecessors, operated by means of a rotary plug. In its simplest form, this consisted of a cylindrical plug or "key" that rotated in a circular socket fixed to the pipe. The plug had a horizontal hole: to turn the water on, the plug was turned (by means of a handle) until the hole was aligned with the bore of the pipe. The water was cut off by giving the plug a quarter turn, so that the hole was crosswise to the flow. Some medieval taps were designed so that the water flowed directly through the horizontal hole in the otherwise solid plug and out a spout at the other side. In other taps the perforation was more complex: upon entering the cylinder, the hole made a vertical turn so that the water either discharged upward out the top of a hollow handle or down out the base of the cylinder.[99]

The taps recovered archaeologically are often fragmentary and come from secondary contexts, so it is seldom possible to reconstruct their exact positions in the water systems to which they belonged. The Lewes tap handle was part of a collection of scrap metal, possibly looted from the nearby Cluniac Priory. The Kirkstall taps were found in fills in the refectory and the kitchen yard, the tap from Westminster Abbey was in a cesspool, and the tap from Saint Gregory's, Canterbury, came out of a backfill layer in the priory kitchen. Some may have been attached directly to pipes, but others are thought to have been fixed to fountain basins, conduit cisterns, or the reservoirs that fed trough lavers. Fortunately, additional evidence for their original contexts is provided by the waterworks plans and documentary sources.[100]

The Canterbury plan apparently uses a small circle with a "pin" through it as a symbol for a tap or stopcock. If this identification is correct, the devices were attached to both basins and pipes in the Christ Church water system. The laver basins are shown with taps projecting from the center of each lobe: presumably water could be drawn off through the taps around the perimeter of the basin. The *Rites of Durham* describes a similar arrangement: the round cloister laver had "many little conduits or spouts of brass with 24 cocks of brass round about it." The plan's purge pipes terminate in similar symbols; presumably these represent drain cocks, which were normally kept closed but which could be opened when it was necessary to empty the settling tanks or pipes for cleaning and maintenance. The same symbols are employed at the ends of short branch pipes that are perpendicular to the main pipelines. These branches are thought to represent standpipes terminating in taps, which allowed users to draw water at various points (such as the kitchen) along the pipeline. The London Charterhouse plan also shows short branch pipes or standpipes terminating in taps, here drawn more realistically.[101]

Taps and stopcocks were made of cast copper alloys. The copper taps could be decorated or gilded: the archaeological examples mentioned above include anthropomorphic, zoomorphic, and floral decorations; the Lewes tap retained traces of gilt. The documentary references to taps (usually in the form of payments to craftsmen) indicate that they were made by various sorts of metalsmiths. At Westminster Palace, Master Robert the goldsmith was paid for making copper taps along with other work on a laver. A copper "key" for a private branch pipe for Otto de Grandson was made by Thomas the plumber. In the mid–fourteenth century, Robert Foundeur was paid for two large bronze taps that supplied hot and cold water to the king's bath. Ramsey Abbey's laver contract with John the Potter called for "sixteen copper keys of subtle design and richly gilt, and fillets through the center."[102]

Rotary plugs are simple to operate, but unlike most modern faucets, they shut off the flow of water very suddenly. Since water is incompressible, the shock waves occasioned by an abrupt cessation of flow can cause water hammer, which may damage pipes. This danger could be eliminated by attaching the taps to tanks instead of the pipes themselves. Many medieval taps do seem to have been mounted in this way, as in the case of taps fitted to laver or conduit cisterns. Where taps were fixed directly to pipes, the provision of intermediate tanks along the pipe run would have reduced water hammer: some suspirals may have served this purpose. It is not at all clear that medieval craftsmen understood the underlying causes of water hammer, but they seem to have devised some effective solutions to the problem.

Rotary plugs are vulnerable to static pressure. The plug is not anchored in its socket: given enough pressure, the cylinder will shoot out of the socket, bursting the tap. Roman and medieval taps shared this weakness. In both cases, attempts to reduce water pressure in the pipelines may have been triggered less by problems with bursting pipes than by problems with bursting taps. Medieval taps were probably even more vulnerable to static pressure than their Roman equivalents. The plugs or cylinders on Roman

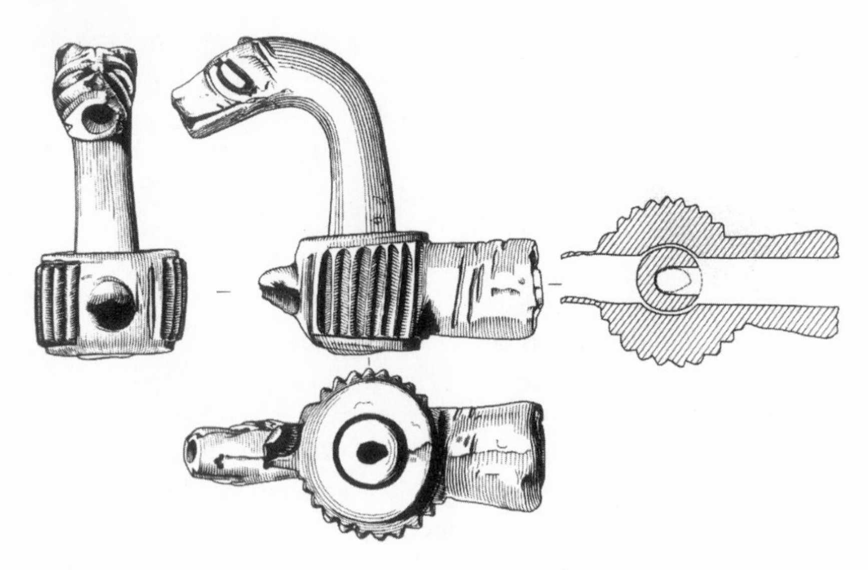

FIG. 3.20. Zoomorphic tap, Fountains Abbey. Two-piece tap, probably associated with the cloister laver of c. 1170. Glyn Coppack, *English Heritage Book of Abbeys and Priories* (London: B. T. Batsford/ English Heritage, 1990), fig. 59A, p. 92.

taps generally have parallel sides: this would have made them rather stiff to turn but also would have helped them resist the static pressure of the water. In contrast, the keys of medieval taps are usually tapered. As a result, any static pressure would tend to lift the plug in its socket. The taps would be easier to turn but more apt to leak and more likely to burst under pressure. Luckily, the same practices that protected against water hammer—fixing taps directly to tanks and providing intermediate tanks on the pipe runs—also helped reduce static pressure.

It is difficult to assess how prevalent problems with taps actually were. Several of the archaeological specimens are keys separated from the main body of the tap. This does not provide direct evidence for bursting taps, but it does suggest that the tapered plugs were fairly easily removed from their sockets. Records of repairs to taps are not very specific, but the replacement of worn or missing plugs may have been the most common problem. The London conduit wardens' accounts for 1335–36 include a payment of £7 11s. 7½d. for the repair of two broken brass "keys," and in 1485 tinker John Clark mended four "watercockes" at Canterbury, at a cost of 3s. 4d. The theft of laver taps by the crowd at Westminster Palace during the coronation ceremonies of Edward II suggests that they were easily pilfered: the thieves probably took the decorated keys but would have had a more difficult time with the main bodies of the taps, which would have been firmly fixed to the laver basin.[103]

WATER SYSTEMS

At present our knowledge of individual hydraulic components is much better than our knowledge of complete water systems. In spite of the valuable contributions of archaeology, the Canterbury and Charterhouse plans still provide the best evidence for the overall configuration of medieval systems. Both have been well studied. Unfortunately, neither is necessarily representative of the usual configurations of monastic (much less urban!) water systems. The Charterhouse layout may have had equivalents at other Carthusian houses, but the requirement of supplying water to individual cells seems to have been unique to that order. The system represented on the Canterbury plan is probably more typical in its general features—most monastic systems fed the laver, the kitchen, and perhaps subsidiary buildings such as the bathhouse, the infirmary, and the brewery—but its design may be more ambitious and complex than that of the average monastic system. When more complete archaeological evidence becomes available for overall system designs, it will be easier to assess the cartographic evidence.[104] Although many systems probably did share the same basic features, there seems little reason to suppose that a single ideal configuration predominated. Every system had to be designed to simultaneously accommodate hydraulic principles, the usually preexisting plans of individual monasteries or towns, and the particularities of local hydrological, topographical, and tenurial conditions.

Many water systems were expanded by adding branches or extending existing supply lines to feed additional discharge structures. A limiting factor in system growth seems to have been the quantity of water. To make up for the restricted quantity of water from a single source, some monasteries and towns eventually ended up with several independent conduits fed by different sources, each serving one or more fountains or other discharge points.[105] Many dealt with the problem of limitations on the quantity of clean water by adopting two-tier water systems: a piped supply of potable springwater was reserved for certain uses, while a more copious supply of subpotable river water was provided by artificial channels.

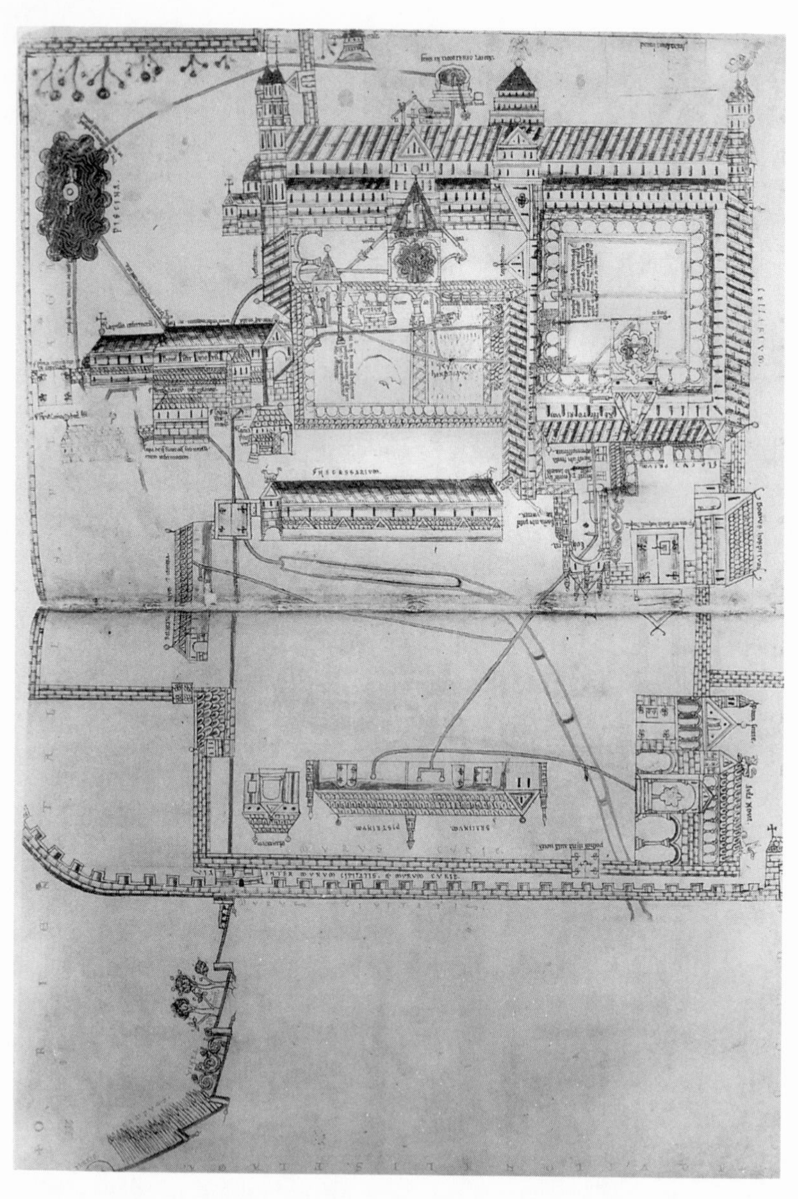

FIG. 3.21. Canterbury plan, showing Prior Wibert's water system (c. 1153–61). Such plans are very rare, but they provide invaluable evidence for the configurations of medieval water systems. Trinity College, Cambridge, MS R.17.1, fols. 284v–285r. Reproduced in R. A. Skelton and P. D. A. Harvey, *Local Maps and Plans from Medieval England* (Oxford: Clarendon Press, 1986), pl. 1A.

The documentary, cartographic, and archaeological evidence do not provide enough information for a full hydraulic analysis of a medieval water system. Future research will undoubtedly lead to a greater understanding of individual components and even the interrelationships of components within systems. Nevertheless, the chances of the recovery and analysis of a complete medieval water system remain low. Our understanding of the parts will probably always exceed our understanding of the whole.

4 Administration and Finance

System administration and finance were two immediate issues facing medieval adopters of hydraulic technology. The construction of a complex water system was a costly enterprise, and once built, it required continuous oversight and maintenance. In monastic communities, various subordinate officials came to be responsible for the maintenance and day-to-day operation of hydraulic installations; some idea of their activities can be obtained from the account rolls recording their expenditures. Municipal governments pursued various financial strategies, often creating new administrative subunits responsible for the water system.

Monastic administration was becoming increasingly complex in the High Middle Ages, and there were differences between orders and individual houses in the numbers and duties of the officials involved. At Battle Abbey, the cellarer was the official in charge of seeing to it that the water pipes were kept soldered and bound. He also paid the cost of cleaning out the latrines and bought two earthen pots for watering the plants in the garden. (Battle was lucky to have had a conduit at all, as the hill on which it was built lacked a good water source. William the Conqueror, however, had insisted that it be built on the spot where Harold had fallen, and he had dismissed the monks' protests with a promise to endow the abbey with a supply of wine "more abundant than that of water in any other great abbey.") Most of the expenditures for repairs to Durham Cathedral Priory's water system fell to the bursar; his account rolls contain numerous payments to workmen for pipe repairs, and he also covered the payments made to the water carriers when the pipes were fractured or frozen, as well as incidental expenses like a new lock for the conduit-head. There seems to have been little systematic maintenance: the payments usually were made during plumbing emergencies, such as the Great Freeze of 1495–96, when eighteen men were hired to help Christopher More de-ice the aqueduct. At Winchester Cathedral, custody of the conduit seems to have been somewhat better regulated. The hordarian (an official in charge of the material resources of the priory) made a customary annual payment of twenty shillings to the warden of the works for maintenance and oversight.[1]

Most maintenance costs seem to have been paid out of the regular revenues of the various monastic officials. These could be supplemented by other sources. Cluny's chamberlain, for example, set aside small monetary gifts for the repair of the pipes of the cloister laver. The costs of construction were sometimes met out of the abbot's own funds, by windfall donations, or by other means. Robert of Scarborough, the dean of York, left Scarborough's Franciscans one hundred marks in his will for the purpose of building a conduit. To pay the legacy, Robert's executor called in an outstanding debt from Meaux Abbey. The monks at Meaux, unable to come up with enough ready cash, were forced to strip the lead off the lay brothers' dormitory and give it to the friars in lieu of seventy-eight marks. Did the friars sell the lead or use it for their new pipes? Boston's Dominicans were helped to cover their construction costs by the bishop, who issued an indulgence for those who helped in the work; the Exeter Cathedral fabric rolls include payments to the scribe who wrote out seven hundred indulgences, sold to help pay for the Cathedral conduit.[2]

The ability of religious houses to maintain their complex water and drainage systems was not always adequate to the task. Even at Canterbury, part of the drainage system could be described as "ancient, ruined and forgotten" by the time Prior Thomas Chillenden (1390–1411) came to repair it. In the early fifteenth century, the Franciscan warden and convent at Southampton were forced to admit that their friary conduit had fallen into serious disrepair, because it had been poorly maintained over a long period of time (which is why they were willing to enter into a joint venture with the city). Religious houses did have the advantage of the threat of spiritual retribution against those who tampered with their systems, however. Civic administrators must have envied Saint John's Hospital in Bristol, which was able to get Bishop John de Drokensford to excommunicate vicar Henry of Aston for breaching their water pipe.[3]

Several cities adopted the simple expedient of placing permanent responsibility for their public water systems in the hands of a single appointed official, often a professional plumber or other hydraulic specialist. In 1376 the city of Bristol drew up a lifetime contract with plumber Hugh White. According to the terms of the agreement, White would not only make and lay new pipes but would also retain responsibility for the maintenance and cleaning of three civic conduits, the Key Pipe, All Saints' conduit, and the conduit of Saint John in Broadstreet, in return for an annual income of ten pounds, to be derived from the rent of certain tenements. The wages paid in 1494 by the stewards of Gloucester to Walter Plummer "in custodiendo aquaeductu" may represent a similar long-term contract with a plumber for conduit upkeep. Later contracts of this sort, retaining local plumbers for conduit maintenance over a term of years, are preserved in the Gloucester records.[4]

Upon completion of the aqueduct for the Fontana Maggiore, the city of Perugia, which had been forced to search throughout Italy for qualified engineers for the project, was determined to keep someone with the necessary expertise on hand to oversee its maintenance. In 1277 the commune made its chief hydraulic engineer, master Boninsegna of Venice, a citizen; the next year it offered him seven hundred lire for a house, as well as a vineyard and other benefactions, in remuneration for his work on the project and to ensure that he would continue to maintain it (at the city's expense) throughout his lifetime. Fra Bevignate, another *incignerius* (engineer) who had supervised much of the project, was given a property near the castle of Sant'Elena for a new Benedictine congregation and seems to have been retained as a kind of communal engineer for some decades.[5]

Some cities developed more complex hydraulic administrations. London instituted official conduit wardens, officials analogous in some respects to the wardens responsible for London's other big public work, London Bridge. They kept financial accounts, collected users' fees, hired workmen for necessary repairs, stored materials and equipment, and generally supervised the use of the conduit. In 1285 a certain Thomas, "Marshal of the Conduit of London," is mentioned; in 1292 four men were selected to be conduit wardens. The fourteenth-century records indicate that normally two wardens served together, apparently for a term of one or two years. In some cases the same individual was reelected. William Hardy, for example, was sworn in as warden in 1310 and again, together with two other men, in 1312. Richard de Gaunt served a brief tenure of a few months in 1325 before being removed from office and replaced; he reappears together with Thomas le Peautrer as warden in the accounts presented for the period running from June 24, 1333, to November 23, 1335.[6]

Keys to the conduit were delivered to the incoming wardens when they were sworn in. The use of these keys is uncertain. The Great Conduit itself may have been locked at night—the conduits at Coventry were fitted with locks so that they could be locked at night and unlocked in the morning. Alternatively, the keys could have been the keys for a storeroom or warehouse where tools, lead, and other materials belonging to the conduit were kept, or they may have been the "keys" of taps or stopcocks. If the latter, their consignment to the new wardens would have served as an actual and symbolic transference of the authority to turn the system off and on.[7]

The outside trades of the wardens are sometimes recorded or suggested by their names: Salomon le Cotiler, John le Coffrer, and Adam le Chaundiler (1292); William le Latoner (1325); Geoffrey de Gedelstone, "cotiller" (1325); Henry de Ware, "isemonger" (1325, 1327–29); Thomas le Peautrer (1333–35); William le Peautrer (1337); Robert le Founder (1350, 1352–53); Arnald Peautrer (1352–53). The high proportion of men from trades that handled nonferrous metals (lattener, founder, pewterer) is striking—presumably their professional expertise stood them in good stead when supervising the maintenance or repair of the lead pipes and other metal components of the conduit. The conduit wardens did not necessarily undertake all maintenance work themselves—like the London Bridge wardens, they could hire outside workers when needed—but their familiarity with metal crafts would have facilitated their ability to assess technical problems, communicate with specialist craftsmen, and purchase materials for the best possible price. The wardens' accounts for 1348–50 indicate that outside workers were hired for some (though not all) conduit projects: three men were hired for three days to bring a pipe into the Mews; five men worked for four days on the repair of a broken *spurgail* (see chapter 3, the section titled "Intermediate Outlets"); and four men were hired for two days to mend and cover a pipe at the fountainhead. The wages paid to these temporary workers were quite high: 8d. per day plus ale for each of the men hired for the first two projects and 6d. per day plus drink for the repair of the fountainhead pipe. Unless the wage inflation in the immediate aftermath of the Black Death was really extortionate, these were probably payments to master craftsmen rather than unskilled laborers. In 1379 responsibility for the provision of unskilled laborers to work on the conduit was assigned to the individual wards—the householders of each ward in turn were obliged to provide a laborer or else work themselves on the conduit or the ditches of the city for one day in five weeks, according to a schedule preserved in the city records; the aldermen were responsible for seeing that the ordinance was carried out.[8]

A change in policy took place in the later fourteenth century. Rather than employing public officials to oversee the conduit, the city farmed out its operation. In 1367 the custody of the conduit together with "its fountain and all its profits and advantages" was leased for a term of ten years to William de Saint Albon, knight, and Robert Godewyn, cutler, with the provision that "the aldermen and sheriffs for the time being may at all times obtain water without payment, and that any of the commonalty may obtain the same, paying for it as of old accustomed." By the terms of the lease, the city received twenty marks' annual rent. The lessees were obliged to keep the conduit in repair above ground, whereas the city remained responsible for any necessary repairs to the subterranean pipes and fountainhead. Whether or not this experiment in leasing out a public utility was successful is not certain—the city records do not indicate what happened when the lease expired.[9]

Viterbo's public fountains fell under the authority of four municipal bailiffs. These officials, one from each quarter of the city, were chosen by the *Consiglio Speciale* and held office for a six-month term. They were responsible for city streets and various aspects of the water supply, including fountain cleaning, and guarded against the misuse of public fountains and watering troughs. Particular emphasis was placed on protecting the water in the basins and the fountain intake systems against pollution. The bailiffs were also responsible for guarding against unlawful diversions of water from the public supply. Private citizens living next to one subterranean water channel, which was fed by the runoff from Fontana Grande, were allowed to lift up the channel's cover slabs in front of their houses and take water for their own use. Otherwise, diverting water from a subterranean channel was prohibited. It was also forbidden to pollute the area around a fountain, though the statutes do not specify whether or not the bailiffs were directly responsible for enforcing this ordinance.[10]

The main administrative mechanism for regulating and maintaining Siena's fountains was fully functioning by 1226. Fonte Branda's basins were under the custody of a municipal fountain warden (*custos fontis Brandi*) whose salary was paid by the city; six other fountains (Vetrice, Val di Montone, Follonica, Ovile, Pescaia, and Foschi) also had their own wardens. It is not known when this practice originated, but it was probably a fairly recent innovation that was linked to the assumption of responsibility for the water supply by the communal government. With the exception of Fonte Foschi, the fountains assigned custodians in 1226 were to provide most of the city's water throughout the thirteenth century and the first half of the fourteenth. The policy of hiring wardens for important civic fountains continued until 1355.[11]

The wardens' term of office was one year (beginning the first of January), and the *Biccherna* (city treasury) paid them annual salaries ranging from four to six lire in two equal installments (generally on June 30 and December 31). The records of these payments have preserved the names of the individual fountain wardens and indicate that the same man (or occasionally woman) could be appointed for more than one term. For example, Siepe di Guido was listed as the warden of Fonte Branda in the Biccherna registers for 1250, 1252, 1257, and 1263; Ildibrandino Ghiandaie, warden of Fonte Branda in 1231 and 1236, reappeared as the warden of Fonte Vetrice in 1249, and in 1251 he was once again warden at Fonte Branda. In the mid–fourteenth century, Fonte Branda was provided with two wardens. For a decade Francesco di Donato and the family of Cinello Dorso (known as Frulla) virtually monopolized the positions. Francesco held the office from 1337 to 1343; Monna Vanna, the wife of Frulla, was warden from 1338 to 1341; Frulla himself held the position from 1341 to 1346; and his sons held it in 1347.[12]

The duties of the fountain wardens were to guard the fountain, its subsidiary basins, and its surrounding area (piazza and streets) by day and night. The main threat to the water seems to have come from people illicitly disposing of filth, defecating, or engaging in other unhygienic activities in or around the fountain. The wardens were supposed to denounce malefactors to the podesta; the city also paid fees to "secret accusers of people who throw filth in public places."[13]

In addition to guarding the fountain against polluters, the Sienese wardens often performed various cleaning and maintenance tasks. They received separate compensation when they undertook such odd jobs, which indicates that these activities were not considered part of their regular duties. Ildibrandino Ghiandaie, while warden at Fonte Branda, received additional payments for emptying the *guazatoio* (a specialized basin of uncertain function) and for general maintenance work. In 1247, when he was not employed as a fountain warden, he was hired to clean Fonte Branda and all its subsidiary basins, and in 1254 he was paid twenty-five lire to clean out the fountain and all its basins after a landslide. Siepe di Guido, in addition to his warden's salary, received payments for cleaning the fountain, emptying out the drainage ditches in the piazza, inspecting the *bottino* (subterranean filtration conduit) for damage, and repairing the road.[14]

Cleaning a Sienese fountain was a cumbersome process. The basins were large, and the job of draining, cleaning, and refilling them was a time-consuming task that rendered a fountain unusable for several days. The Biccherna records for 1307 indicate that it took two days to clean a single basin, five days to clean a fountain and *abbeveratoio* (watering trough for animals), and seven days to clean the fountain and three subsidiary basins at Fonte Branda. Minor repairs to the basins were most easily effected while they were empty. In 1284, for example, Uberto Saraceni was paid for cleaning the fountain, abbeveratoio, and *lavatoio* (laundry trough) of Fonte d'Ovile, and at the same time he received compensation for repairs to the masonry of the lavatoio apertures. The recorded intervals between cleanings vary, but in general cleaning seems to have been an annual task performed during the late spring or early summer: most payments fall within in the latter part of the first half of the fiscal year (which ended June 30). Usually all the basins seem to have been cleaned at the same time; since they were interconnected, it would have been easier to drain and refill them in one operation. Extra guards were sometime hired during the period when the fountains were empty to ensure that women would not draw water there—the women may have been tempted to walk across the empty basin and draw water directly at the spout, an abuse that would not only dirty the basin but could damage its lining. The municipal authorities recognized that the withdrawal of a public fountain from use during the cleaning process posed a problem that might be ameliorated through technological innovation. There were various calls for modifications to basin designs to facilitate emptying and to eliminate the problem of stagnant water, although it is unlikely that these resolutions led to fully satisfactory technological solutions.[15]

Until the end of the thirteenth century, the city guarded, cleaned, and maintained the fountains on an individual basis, hiring a fountain warden and other workmen as necessary, but with little crossover of personnel between fountains. The city maintained an intensive involvement, as expressed in the employment of a full-time fountain warden and a high level of expenditures for cleaning, maintenance, repairs, and structural modifications, in a fairly stable group of approximately eight fountains. The prominence of these fountains in the documentary and architectural record has tended to skew scholarly discussions of Sienese fountains, although some two dozen other fountains in the city or its rural hinterland are known to have existed in the thirteenth and the first half of the fourteenth century. The latter group of fountains received less civic support and not much is known about them. Though some did have subsidiary features, they seem to have been structurally less complex and less homogenous than the better-known, centrally administered main fountains. The city administration occasionally appointed wardens at a few of these lesser fountains, but this experimental practice never developed into a consistent policy. The thirteenth- and early-fourteenth-century commune seems to have had a two-tiered fountain policy: it owned and invested heavily in a few large fountains, which were provided with a standardized set of subsidiary features; a more numerous and more heterogenous group of fountains seem to have remained outside the direct control of the central administration and received only sporadic governmental support.

Until the mid–fourteenth century, Sienese fountains fell under the rather general supervision of officials in charge of roads, bridges, and fountains, but there was no organizational unit or official solely responsible for fountains. Important decisions concerning fountain projects were made by the *Consiglio Generale,* whereas the Biccherna paid the salaries of individual wardens and master workmen. Toward the end of the century the beginnings of an organizational restructuring can be seen. The administration of fountains became more centralized, and specialist workmen became responsible for multiple fountains. The tendency first becomes apparent with fountain cleaning. In 1281 an experimental step in the direction of greater centralization was taken when Albertuccio Ranerii was hired to clean four fountains. In 1298 Ceccho Amate was paid to clean four fountains together with their subsidiary basins, and by 1307 Pietro Schotti was cleaning seven civic fountains. Pietro Schotti and Cienino di Fino were officially responsible for cleaning the fountains in two of the three city *terzi* in 1309; they were paid for ten fountains that year. At the upper administrative levels, the *maggior sindaco* of the commune obtained general jurisdiction over roads, bridges, and fountains in 1319. According to the constitution of 1337–39, he was to ensure that all officials, individuals, and committees responsible for urban hygiene, public construction, and the maintenance of these public works performed their functions satisfactorily and was to conduct regular tours of inspection of all the fountains and bottini. In 1321 a certain Lemmo was paid twenty-five lire for the cleaning and maintenance of (all?) the city's fountains; in 1329 he received a salary of fifteen lire for similar duties at eight fountains.[16]

Ceffo Venture, a master mason, appears as the official and workman in charge of fountain maintenance in 1335. By 1337 both Lemmo and Ceffo held official appointments as fountain workmen. Ceffo maintained his position as the master and official in charge of the city fountains, alone or with a partner, until 1348, at an annual salary of 40 lire. As such, he was responsible for fountain and bottino inspections, maintenance, repairs, and cleaning and was in charge of overseeing at least some of the fountain wardens. He also kept accounts of purchases of materials and other expenses incurred in fountain projects. Ceffo's own detailed accounts have not been preserved except in brief summaries, but account books of later fountain officials contain a wealth of information on tool inventories, materials, and operational expenditures. The Biccherna registers indicate that Ceffo worked on at least fifteen different fountains. The only important exception was the Campo fountain, which was under the separate jurisdiction of three special officials.[17]

The employment of an official like Ceffo Venture, directly responsible for the city's fountains, seems to have led to a greater number of fountains being brought under more direct civic patronage. As a result of the new centralization, the Biccherna accounts tend to record combined payments for works on multiple fountains, without always specifying which individual fountains were included. In consequence, the aggregate numbers of individual fountain citations in the fourteenth-century records show great fluctuations from year to year, which tend to obscure more general policy trends. In years in which the individual fountains are specified in the records, however, the overall numbers are higher than in the preceding century, ranging from thirteen to fifteen, and include a broader range of fountains. Ceffo, for example, frequently undertook works on the minor as well as the main fountains. By creating a more streamlined, centralized administrative unit directly in charge of workmen, materials, and multiple fountains, Siena was able to muster a flexible and effective response to the maintenance requirements of an increasingly complex water system. Nor was Siena the only city to develop a sophisticated hydraulic administration. The master builder for Nuremberg, for example, kept careful and thorough records, which note in detail the exact locations and depths of all underground components, so that the pipes and fountains could be kept clean and in good working order.[18]

Communities had to find adequate ways to finance the construction, oversight, and maintenance of their water systems. Complex hydraulic systems were expensive to build. Operating costs (including the salaries of wardens and workmen and the materials needed for repairs, cleaning, and general maintenance) were less onerous but were unremitting: a failure to meet maintenance costs could lead to the breakdown of the entire system. Municipal officials adopted various financial strategies to pay for their water systems.

Some cities used general public revenues; others employed user fees, occasional levies, or other extraordinary sources of income. Siena's water system was financed with public revenues. Variations in the sources of income used to build and maintain the bottini and fountains tended to follow more general trends in the government's search for public revenues, and water system finance was subsumed in the overall fiscal policies of the commune. The bulk of Siena's income was obtained from a combination of taxes and public loans. The surviving Biccherna volumes preserve a detailed (if incomplete) record of the commune's incomes and expenditures, including multifarious expenditures on public fountains and the bottini network. In addition to utilizing its usual sources of revenue to fund the public water system, the commune did levy a special assessment for fountain, road, and bridge maintenance on the communities in the Sienese *contado.* Siena occasionally sought to force the Vescovado (those lands directly subject to the Sienese bishopric) to pay for the maintenance of public fountains, roads, and bridges, and in a statute of 1295, the commune ordered churches and monasteries to pay taxes for the maintenance of these public works. This policy of clerical taxation met with resistance inasmuch as it was considered to be a threat to the liberty of the church: when the clergy could be forced to pay, they seem to have preferred to call their payments voluntary donations.[19]

From the late thirteenth century on, *gabelles* (indirect and excise taxes) represented an increasing proportion of the city's revenue, which led to a corresponding rise in the influence of the office that administered this source of income, the General Gabella. The general trend toward an increased dependence on gabelles as a source of civic revenue was reflected in the ways money was raised for hydraulic projects from the mid–fourteenth century onward. The gabelle on the city of Grosseto (which rendered some 1,000 lire annually) served as a major source of funding for the ambitious Campo fountain project from 1341 to 1347. In the early fifteenth century the income from the gabelle on bread was assigned to the maintenance of the water system.[20]

Although nearly everything else in Siena seemed to be subject to taxation in one form or another, the public fountains continued to supply water free of charge—the city was willing to shoulder the financial burden of providing such an essential public service. Siena's expenditures on its hydraulic infrastructure stimulated the city's economy—a point that was appreciated at the time, particularly in regard to the demands of the preeminent wool industry. Conversely, the commune's ability to build and maintain its water system was directly dependent on the prosperity of its citizens, the efficiency of the financial magistracies and their ability to generate public revenues, and public support, for expenditures on the urban infrastructure.

The yearly sums spent on the water system show considerable fluctuations. Salaries for fountain wardens were a modest, predictable expense, and the costs of normal maintenance, repairs, and cleaning (although varying from year to year) remained moderate. In 1307 (a year in which ten fountains were emptied and cleaned), a total of 12 lire was spent on wardens' salaries; cleaning costs (including salaries) came to 35 lire 11s. In 1341 salaries and maintenance costs together reached 63 lire 19s. Major renovations and new construction projects, however, were far more expensive. Bottino works (including salaries for masters and workmen) cost 206 lire in 1307; in 1334, 6,000 florins were allocated for work on the new Campo fountain, a project that would end up costing the city considerably more than its initial estimate. Siena's water system was not cheap, but it does not seem to have been a disproportionate drain on the city's finances. Even in 1307 and 1341, years of unusually high hydraulic expenditures, such payments constituted considerably less than half of 1 percent of the Biccherna's overall expenditures. The financing of Siena's water system was an impressive achievement. The council was generous in its support for new hydraulic projects, and the tedious burden of paying for cleaning and maintenance was dutifully assumed by all factions in the ruling hierarchy throughout the communal period. It was an achievement that was linked, however, to the overall economic health of the community. In the later fifteenth and sixteenth centuries, the city seemed unable or unwilling to meet the constant financial demands of system maintenance; thus, expenditures declined and the public water system fell into a state of decay.[21]

The construction of London's Great Conduit was financed in part by a contribution in 1237 of £100 "to the water conduit" from the merchants of Amiens, Corby, and Nesle, who received certain trading privileges in exchange. The London conduit wardens were responsible for collecting the money left to or acquired by the conduit, expending it in conduit maintenance and rendering an account of their expenditures for auditing when required to do so. Unfortunately, the original financial records do not survive, but the city Letter-Books occasionally preserve more than the mere notation that the audit has taken place. The wardens collected user fees from brewers, cooks, and fishmongers, which helped finance conduit maintenance and repairs, although London's periodic reversals in policy toward these users rendered this an inconsistent source of revenue. A summary of the wardens' incomes and expenditures for the period 1333–35 indicates that the annual income received in quitrent for tankards and tynes (a tyne was a wooden staved bucket resembling a barrel with a handle) was fairly predictable: £6 18d. (June 24, 1333–June 23, 1334); £6 6s. 6d. (June 24, 1334–June 23, 1335); £6 13s. 4d. (June 24–November 23,

1335). The fact that the income for the first five months of the final term is equivalent to the annual incomes of the previous terms suggests that the charges assigned to users for tankards were collected in one annual sum, toward the beginning of the fiscal term. Expenditures were more varied. In 1333–34 the wardens spent £9 10s. 2d. for lead and other necessities; in 1334–35, £4 22d. was spent to make a clay wall around the conduit-head at Tyburn; in the second half of 1335, the repair of two broken brass taps (*clavorum*) cost £7 11s. 7½d. Unfortunately, the figures preserved in these last accounts are somewhat suspect, since Richard de Gaunt, one of the two wardens, seems to have been guilty of financial impropriety. In 1337, after his term of office had expired, he was convicted of embezzling lead and money belonging to the conduit.[22]

In the accounts for a two-year period ending in 1350, the income for London's conduit was somewhat higher. A total of £15 13s. was taken in, in spite of the impact of the Black Death, which reduced the second year's total. In this two-year period, payments for ordinary operating costs came to £6 5s. 2d. These included the regularly scheduled cleaning of the fountainhead; repairs to the fountainhead, two *spurgails,* and two sections of pipe; workmen's wages and ale allowance; the hire of a cart; opening and closing the conduit; the hire of the two "vadlets [servants]" to collect the money for the tankards; a year's hire of a house to store the tankards; and two irons used for stamping tankards. More unusual expenses incurred during the same period were the cost of laying a new pipe to the Mews (6s. 6d.), the purchase of a fozer (fother?) of lead (8 marks, 12d.), and the cost of examining the conduit at the mayor's command when it was "slandered for poison" (32s. 2d.). Overall, these figures indicate that expenditures for normal repairs and maintenance could be covered by the incomes received, especially if surplus incomes from years of low expenditure could be set against the occasional deficits arising in years of higher-than-average expenses. Major repairs to the system or system growth would have been less easily accommodated. The conduit was not completely self-sustaining but depended on periodic infusions of outside money, such as testamentary benefactions or special city measures in times of unusual expenditure.[23]

In 1378 the London authorities resorted to a form of not-so-gentle persuasion to raise money for some needed conduit repairs. They resolved to summon the good men of each ward to appear in the Guildhall, where they would be urged "to make a free gift according to their wealth and zeal for the city's welfare." Those who "maliciously" refused this appeal to their better natures were to have an assessment made of their wealth. In 1445 a levy of a fourth part of a fifteenth in each ward was raised for use on the common aqueduct, whereas a fine collected from Sheriff Robert Byfeld (found guilty of using "unfitting wordes" in an argument with the mayor) was applied to conduit repairs in 1479.[24]

London's collection of conduit revenues through levies and fines found a parallel in fifteenth-century Coventry, where modest quarterly payments were levied on the inhabitants of the wards "towards the repair of the conduits." Those who evaded the tax were to be distrained for double the amount. Fines against brewers and other tradesmen who illicitly fetched water at the conduits in connection with their trades and against property owners who failed to clean the river were also applied to the maintenance of the town's conduit.[25]

Neighborhood fountains were frequently financed wholly or in part by the residents of the ward. In 1390 the men of London's Farringdon Ward sought permission to lay a branch pipe from the Great Conduit to supply their ward "at their own cost and charges." Likewise, the inhabitants of the parish of Saint Giles without Cripplegate had built a cistern "at great cost" in 1483 to receive the water supplied by pipes from Highbury, which had been paid for by the executors of the late William Estfeld. The residents petitioned the Common Council to vest the cistern and water in them in perpetuity, "subject to the right of every inhabitant of the city to take water from the said cistern at will, and that the repair of the cistern, pipes, &c., should be made at the cost of the city, as in the case of other cisterns and conduits." Viterbo permitted the residents of its *contrade* (wards) to build fountains as long as they paid the costs themselves. In Siena the cost of a late-fourteenth-century attempt to supplement Fonte Branda's water supply (which fed various industrial pools and mills) was split between the beneficiaries: the commune, the Arte della Lana, and mill owners.[26]

The lands and water sources for urban water systems were occasionally obtained by donations from private owners. Similarly, private charitable bequests for public works sometimes served as a source of revenue for urban water systems. In the early Middle Ages, motivations for water patronage seem to have shifted away from public displays of munificence, designed to benefit and impress one's fellow citizens, toward a more Christian impulse of charity, helping the poor. Some late medieval water projects still followed this model. Thomas Knolles, for example, arranged to supply the poor prisoners in Newgate and Ludgate by means of lead pipes, which conveyed the excess water from the Hospital of Saint Bartholomew's cistern.[27]

Most benefactions for public works benefited the entire community, not just the poor. Donations to public works that served all citizens, such as conduits, bridges, and even public latrines, emerged as a type of charity in English wills in the thirteenth century and became increasingly common in the fourteenth and fifteenth centuries. In effect these were outright benefactions to the community as a whole. In some respects they recall the tradition of Roman civic patronage, but they retained the conceptual framework of Christian charity. Bequests for public works are embedded among more traditional forms of pious donations, such as bequests to friars, churches, chantries, anchorites, and poor prisoners.[28]

The appearance of these civic benefactions among more traditional charitable bequests may be linked to a gradual shift in thirteenth- and fourteenth-century concepts of charity. Testamentary bequests as a restitution for sin were supplemented and to some degree supplanted by a new, activist style of preemptive philanthropy.[29] Not all benefactions were necessarily pious, of course. Civic pride, personal prestige, the desire for commemoration after death, and political ambition could inspire medieval as well as Roman patrons to endow public works. Whatever the motives, the growing popularity of these civic benefactions helped underwrite the construction and maintenance costs of several urban water systems. In some cases the sudden windfall of a private benefaction actually triggered the initial adoption of a new system.

In Viterbo members of the powerful Gatti family were particularly active patrons of public works, commemorating their sponsorship in a number of inscriptions composed in leonine verse. Visconte Gatto sponsored an aqueduct in 1268 to feed a fountain in the papal palace built by his father, Rainerius, two years earlier. The Gatti coat of arms decorates the neighborhood fountain of S. Carluccio and the palace of the captain of the people, and Gatti inscriptions record their patronage of the city wall (1268), the hospital Domus Dei (1303), and the tower of S. Biele. Gatti patronage seems to have been linked with the clan's political ambitions. Several members of the family wielded power by means of civic magistracies, and they seem to have employed a deliberate policy of using their highly visible patronage of public works to bolster the family's popularity and prestige.[30]

Sponsoring of public works during the patron's lifetime occurred in Ireland also. In 1308 Dublin's mayor, John le Decer, known for his generous benefactions to the city's friars and his patronage of other public works, such as the construction of a bridge over the Liffey, erected at his own expense a marble cistern to receive water from the city's conduit.[31] More typically, however, such benefactions were made in the form of testamentary bequests; whatever other motives impelled the sponsors, by the time their projects were implemented, their political ambitions had been laid to rest.

The restoration and transfer of the friary conduit to the town of Southampton were occasioned by a testamentary bequest made by former mayor John Benet, who left money to the town for this purpose. According to the transfer deed, the friary enacted the agreement "to the honor of God and for the health of the soul of John Bennett." Benet had probably negotiated the terms with the friars before he died. Frequently the donors were, like Benet, former mayors. Simon Grendon, who left twenty pounds to construct a conduit bringing water to the *quadrivium* (crossroads) in Exeter if the project was completed within four years of his death, had been mayor of Exeter three times. Fourteenth- and fifteenth-century testamentary benefactions by former mayors such as Adam Fraunceys, John Philipot, and William Estfeld financed most of the new extensions to London's water system. Mayors were not the only benefactors, however. Contributors to London's conduit included Katherine la Fraunceyse; William Love, "fourbour" (furbisher); draper John Gille; Edith, the widow of brewer Simon Derlyng; merchant John Leycestre; vintner John Walworth; and girdler John Costyn. Some of these bequests were applied to "work on the conduits" generally, whereas others were earmarked for particular projects or maintenance of the system.[32]

A complex urban water system could prove to be a substantial financial burden, and inadequate financing could be disastrous for the system. The early modern period witnessed the decay of Siena's bottino network and the eventual abandonment of many of fountains, but the seeds of the impending problems were sown when payments for the water system's maintenance were cut back in the fifteenth century. In England, Hull's conduit, for which the city had obtained a license in 1447, proved short-lived because of an embarrassing financial crisis. By April 1449 at least thirty-six pounds had been spent on the project, but in 1461 the town was forced to dig the lead pipes up again and sell them off to pay its debts. This abrupt shift of policy does not seem to reflect any dissatisfaction with the technology itself, since there was an attempt to revive the conduit some years later.[33]

Medieval decisions to reject potential urban water systems are not easy to document, but most monasteries and cities did not adopt complex systems. Water systems were costly. Although they could provide long-term benefits to the community, they were substantial and immediate drains on municipal and ecclesiastical budgets. The communities that did adopt them frequently demonstrated an innovative flair when it came to financing their construction and maintenance. Ultimately, the adoption and continued success of complex systems rested on the willingness of the community to bear its costs, whether through increased levels of taxation and expenditures or through the generosity of private individuals.

5 Users

The successful adoption of a new water system by its intended users does not necessarily follow its initial construction, as setbacks to twentieth-century attempts to introduce modern plumbing into traditional communities have demonstrated.[1] In the case of medieval monasteries, where the new water systems were designed to accommodate existing and closely regulated patterns of water usage, the monks seem to have had few problems adapting to the new technology. When medieval cities adopted public water systems, however, the results were less predictable. Urban inhabitants' perceptions of the new fountains determined whether or not they decided to transfer their traditional patterns of water usage to them. For all their apparent advantages, fountains also had potential disadvantages. Compared to flowing rivers, they supplied far less water, the frontage area for user access was more restricted, and the smaller body of standing water was more vulnerable to pollution.

Even when urban dwellers did decide to use the new fountains, their transferred behavior was not always technologically appropriate. The technological diffusion of complex water systems necessitated a contingent social diffusion of appropriate patterns of behavior. The particular activities that could be appropriately (or inappropriately) transferred to the new systems depended on the design of the distribution structure. British conduits were generally closed cisterns fitted with taps, a design that facilitated filling vessels but was not well adapted to other patterns of water use. The closed design had the advantage of protecting conduit water against pollution by users. On the Continent, fountains were often designed with open basins. Such an open body of water made it possible to transfer a wider range of activities to the fountains. The fountains may have been originally intended as a supply of drinking and domestic water, but people quickly discovered that they could also water animals, bathe, wash clothes, or carry out industrial activities in the convenient new water sources. Such activities, however, were inappropriate uses of a single basin, since they rendered the remaining water unfit for consumption. Civic authorities tried a variety of expedients to protect their fountains against misuse: statutes, guards, informers, and fines. The informal development of new behavioral norms and unofficial resolutions of disputes among users themselves were probably as important as official interventions in the establishment of appropriate patterns of usage. Most effective in the final analysis, however, was the adaptation of the technology to accommodate the needs of competing users.

The provision of water for drinking and domestic use was a primary function of most medieval fountains. Medieval attitudes toward water as a beverage were not entirely negative, but in general other drinks (especially ale, beer, or wine mixed with water) were preferred. The response of the boy-monk in Aelfric's dialogue, when asked what he drank, could probably represent the great majority of northern Europeans: "Beer if I have it, or water, if I don't have beer." The *Rule of St. Benedict* had conceded that monks could not be expected to drink water alone; and by the High Middle Ages, the monastic allocation of beer and wine seems to have been exceedingly generous. To the extent that they could afford it, the laity did not lag far behind. Thanks to the strong demand for beer and ale, brewing came to be one of the more popular occupations for women, in both villages and cities. Eldreth, the wife of Ingenulph (the plumber of Christ Church,

Canterbury), sold ale to the monks to the value of eight pounds, a figure four times her husband's annual salary.[2]

These cultural habits and attitudes were based at least in part on the health risks associated with drinking untreated water. A poem by Hugh Primas takes the form of a dispute between wine and water. One of the charges leveled against water relates to its unfortunate gastric effects:

> Who has of you mistakenly partaken
> Finds himself of health forsaken
> With sudden violence
> His belly rumbles by wind inflated,
> And, courteously restrained, undisseminated,
> Oh, the pain intense.
> Yet the belly constricted by such distress
> Finds its relief, release nonetheless
> through the lower throat.[3]

Water did have its defenders. According to the Franciscan chronicler Salimbene, when Joachim of Fiore was discovered to be secretly drinking water rather than wine, he praised water for being a temperate drink that did not inebriate the drinker or loosen the tongue. Nevertheless, this view seems to have been a minority opinion: even the holy Joachim was drinking water only because he had incurred the rancor and malice of the monk in charge of the refectory.[4]

Drinking water was traditionally drawn from natural outlets such as rivers and springs or from wells. Hildegard, abbess of Bingen, considered well water better to drink than springwater, which was itself superior to rainwater and river water. Snow water was dangerous to the health, whereas river water and swamp water should always be boiled and then cooled before drinking. Of course, boiling water before drinking it was not always a practical option, and given the extra costs in fuel, time, and labor, it is doubtful that Hildegard's sensible advice was widely followed. Saint Francesca Romana and her friend Vannozza both fell into the Tiber when bending for a drink: the author of Francesca's *vita* (life) considers their escape from drowning a miracle, though the fact that they did not die from the polluted water is perhaps equally astonishing. Travelers were at a particular disadvantage, not knowing which local sources offered safe drinking water. The author of the guidebook for pilgrims to Compostela devotes a chapter to the discussion of which rivers along the route had potable water and which did not. The pilgrims were warned against trusting local residents, especially those to be found along the Rio Salado, who seem to have made quite a tidy living out of its deadly waters. "While we were proceeding towards Santiago, we found two Navarrese seated on its banks and sharpening their knives: they make a habit of skinning the mounts of the pilgrims that drink from the water and die. To our questions they answered with a lie saying that the water was indeed healthy and drinkable. Accordingly, we watered our horses in the stream, and had no sooner done so, than two of them died; these the men skinned on the spot." In York (a city that did not build a medieval water system), archaeological evidence indicates that environmentally sensitive species of fish had disappeared from the Ouse by the tenth century. A comparison of human skeletons from Saint Helen-on-the-Walls's cemetery with those from the cemetery at rural Wharram Percy shows that York's residents were much more likely to suffer from anemia, probably as a result of endemic diarrhea caused by ingesting unclean water.[5]

Wells could be lethal booby traps for the unwary. English coroners' rolls reveal that children and women were plummeting down village wells with alarming frequency. Many of these rural wells may not have had much, if anything, in the way of safety barriers or markers. A commission sent to inquire into a dangerous well in the high road from Egham to Staines found that when the road was flooded, the well could not be seen, and men and cattle were plunging to their deaths. The commissioners discovered that the abbot of Chertsey was responsible not only for this well but also for several other perilous wells in the area and that two unknown men had drowned in them within the past two years. (The flooding in the road, which rendered the unmarked wells invisible, was attributed to the same abbot's neglect of the riverbank.) The abbot does not seem to have been unduly perturbed that, thanks to his negligence, "none could ride, drive a cart or go on foot without great peril of life" through his lands, but he did make sure to claim the money found in the purse of one of the victims. Urban wells seem to have been safer than rural ones: perhaps they were provided with more substantial wellheads. The fatal accidents associated with London's wells sometimes involved workmen who had intentionally descended into the shaft, such as John de Maldone, who was overcome by foul air when cleaning an empty well, or John Bone, who fell and drowned while climbing down a well by means of a long pole to retrieve a bucket. Compared to rivers and wells, conduits and fountains were safe: the water they provided was usually wholesome, and users did not approach them in peril of their lives. Fatal accidents associated with complex water systems generally involved workmen, not users, and they do not seem to have been common.[6]

The provision of civic water systems probably led to a considerable (but not staggering) increase in domestic water consumption on the part of urban residents. Users would draw water at the fountains and carry it away in some sort of vessel. In traditional societies, women and children, especially girls, are often the family members primarily responsible for fetching domestic water. They usually go for water first thing in the morning and repeat the trip several times during the day. In medieval Europe, too, drawing domestic water was generally (though not exclusively) women's work. Drawing water at the fountain played a social as well as a functional role in women's lives. Even in a relatively gender-segregated society like medieval Italy, it created an opportunity for respectable women and girls to come out into the public world of the piazza and interact with their women friends. It also offered women the chance to discreetly observe—and to be observed by—males outside the limits of their immediate family group.[7]

A popular tradition concerning a broken water jar links Viterbo's civic saint, Santa Rosa (1233–51), with the fountain that stood near her house. The young Rosa had gone to draw water with some other girls. In the process, one of her companions broke the vessel she was carrying, shattering it into many pieces. The girl, fearing her mother's scolding, falsely blamed Rosa for the accident, whereupon the saint caused the broken fragments to reassemble into a complete, undamaged vessel.[8] The tradition may be legendary, but the scene it portrays of a group of girls drawing water together at a fountain must have been a common sight in thirteenth-century Viterbo.

The basic Viterbese fountain type has a large, circular, open basin with a shaft rising from the center. The water rises in a pipe concealed in the core of the shaft and discharges into a small, hidden basin midway up. The water then flows out through projecting spouts and falls into the large open basin below. Water is collected either by dipping a vessel in the open basin or by holding the vessel under one of the spouts. To make the job easier, users can rest the base of the water jar on a stone block situated in the main basin below each spout. Supporting ceramic water jars in this way not only eases the strain on the user's arms but also puts less stress on the vessel handle. Judging by the abundant archaeological remains of fractured water jars, the accident that befell Rosa's friend was not unusual. The typical medieval pitcher fracture shows a failure of tensile strength, "the neck and handle being literally pulled away from the body of the jug," and was probably caused by the sudden lifting of a full vessel without supporting the base.[9]

In Sienese fountains, the main basin was reserved for drawing water for drinking and domestic use. Water from the *bottino* (subterranean filtration conduit) flowed into the basin from an inaccessible spout at the back, so users had to dip their vessels into the standing body of water. Catherine of Siena alluded to this Sienese practice of immersing a water jar in the fountain in a letter to Monna Melina, in a complex metaphor that blends biblical allusions with the daily experience of a Sienese woman: "So God insists that we bring with us the vessel of our free will, with a thirst and willingness to love. Let's go, then, to the fountain of God's sweet goodness. There we shall discover the knowledge of ourselves and of God. And when we dip our vessel in, we shall draw out the water of divine grace, powerful enough to give us everlasting life." To protect the purity of the supply, women were required to wash their water jars in a separate basin before dipping them in the fountain.[10]

Male servants and apprentices also carried water. In 1300 London's Geoffrey de la March was carrying water in a tyne (a wooden staved bucket resembling a barrel with a handle) for the use of his master. As he was carrying the heavy tyne from Ludgate, he was accosted by Adam de Hidecroun, who took the water and filled his own pot with it. Geoffrey started cursing, whereupon Adam grabbed Geoffrey's stick and hit him on the head, upsetting the tyne and spilling the rest of the water in the process.[11]

Both women and men worked as professional water carriers. Durham Abbey's monks were periodically forced to hire water carriers (often women) to carry water from the Wear River to the abbey's brewhouse, bakehouse, and kitchen when the conduit pipes were fractured or frozen. The 1496 guild regulations for the London water-bearers were addressed to both brothers and "systers" of the fraternity, who generally used large tankards or tynes rather than earthenware vessels. Tankards, like tynes, were made of hooped staves, but tankards were cone-shaped. They had a small iron handle at the narrow (bottom) end, were fitted with a bung, contained about three gallons, and were carried directly on the shoulder. Water carriers also employed carts to haul larger casks of water. In Rome, members of the Compagnia degli Acquariciarii transported their water in barrels strapped to the backs of donkeys.[12]

An analysis of the Sienese 1285 direct tax (*dazio*) returns lists ten water carriers, all women, who were among the poorest class of those liable for taxation. English water carriers were not necessarily so impoverished. The lay poll tax returns from York show that Richard Waterleder paid 18d., which was well within the range of rates paid by other townsmen and was higher than laborers' rates (overall the sums paid ranged from 4d. to 20s.). Some were even quite prosperous. In 1348 "waterlader" Geoffrey Penthogg of London left his son and his wife two gardens, one messuage, "and also another." In 1349 his wife Johanna left, in turn, five horses, two carts, the tenement bequeathed to her by her late husband, and one portion of hay to be sold for pious uses; in addition she left a house and a garden to her brother.[13]

British conduits tended to be enclosed structures. London's Great Conduit was a "cesterne of leade castellated with stone." Water was released from the lead cistern through pipes, which may have been fitted with taps, into a square stone basin. Users filled their vessels by holding them under one of the pipes. Consequently, the number of simultaneous users would have been limited to the number of available outlets.[14]

London's conduit water proved so popular that bitter and long-running disputes arose between domestic users and the tradesmen (especially brewers) who also drew water there. The conflicts centered on two issues: accusations that tradesmen were wasting water (perhaps indicating that the total supply fell short of the total demand) and quarrels over access to the conduit (indicating that the design of the distribution structure failed to adequately meet the needs of multiple simultaneous users). Behind the complaints lay the assumption that domestic users had first rights when it came to conduit water. According to a complaint against the brewers filed in 1345, "of old a certain conduit was built in the midst of the city of London, so that the rich and middling persons therein might have water for preparing their food, and the poor for their drink." This perception of the conduit's main function was shared by civic officials, who tried a variety of expedients to deal with the situation. Initially their provisions were ad hoc responses to specific complaints rather than long-term attempts to address the underlying structural problems. On several occasions they banned particular types of users, such as brewers and fishmongers, from the conduit, but such blanket prohibitions proved untenable and seem to have been reversed as often as they were imposed. In the second half of the fourteenth century, the city government took steps to improve the system by increasing the total supply of water and by providing new distribution points to accommodate more users.[15]

Conflicts between Londoners over conduit water are not recorded for the later fourteenth century. This may simply indicate that the old complaints were being heard by the new private operators rather than the municipal authorities, but it is possible that a greater equilibrium between supply and demand had been achieved. The problem of insufficient water would have been alleviated by the drop in population following the onslaught of the Black Death. (Of the ten houses charged for water in the two years recorded in the November 1350 wardens' account, three were described as having been "one year empty." Their owners had presumably either died from the plague or fled the city.) Furthermore, the amount of water delivered to the system had increased. In 1355 an additional source of water had been obtained by Alice Chobham's grant of a spring in Tyburn. In the late fourteenth century, the system was expanded by the addition of two new distribution points. In 1378 a common council considered the best means of repairing the conduit in Chepe. This resulted in an eastward extension of the pipeline to Cornhill, where an existing stone prison called the Tunne was converted into a water cistern. In 1389 another conduit, the Little Conduit, was built.[16]

Neither of these new distribution points increased the total amount of available water, since they both derived their supply from the same pipe system as the Great Conduit. They did, however, make piped water more accessible to neighborhoods to the east and the west. In the course of the fifteenth century, more new conduits were built at Paul's Gate, Aldermanbury, Fleet Street, Cripplegate, Gracechurch Street, and Oldborne, and by the sixteenth century London had nine conduits or water bosses in the area west of the Walbrook.[17] The creation of new distribution points made collecting water more convenient for citizens residing near the new conduits; since the water supply was now available from multiple cisterns, the net result would have been an increase in the number of users who could obtain water simultaneously. The increased capacity to accommodate multiple users, however, may have been offset by an increase in the total number of users once conduit water was conveniently available to citizens in a wider catchment area.

Though the expansion of London's supply system and the addition of new outlets may have temporarily alleviated the shortage of water, they did not completely solve the problem of overcrowding at the Great Conduit. In 1415 renewed complaints against the brewers resulted in yet another ordinance. The brewers were permitted to continue renting "the fountains and great upper pipe," but the small lower pipes were reserved to the use of the common people.[18] This time instead of issuing blanket prohibitions against particular categories of users, the municipal authorities attempted to accommodate the needs of all by reserving specific parts of the structure for specific users. This may have eased tensions somewhat, but it does not seem to have solved the fundamental problem of overcrowding. A satirical verse, accompanying a print entitled "Tittle Tattle," reveals that the tensions that arose from overcrowding had not significantly abated in the Elizabethan period:

> At the conduit striving for their turn
> The quarrel it grows great,
> That up in arms they are at last,
> And one another beat.[19]

Other cities suffered similar problems. In King's Lynn, crowds jostling to fill their vessels were damaging the conduit arches and stonework. The city authorities established a policy of first come, first served, regardless of social class, and placed restrictions on filling large vessels. Those who cut in at the head of the line were fined—it is tempting to speculate that the British passion for queuing may have originated in such lines of citizens forced to wait patiently (or not so patiently) for their turn at the local conduit. Coventry's conduit also triggered conflicts between domestic and industrial users. The fifteenth-century Leet Books contain repeated injunctions against industrial users of conduit water. Brewers or maltsters were not to fetch water at the conduit for brewing or steeping (although they were permitted to draw water there to prepare food). As in London, outright bans seem to have been difficult to maintain. In 1493 those that

FIG. 5.1. "Tittle-Tattle; or, The Several Branches of Gossipping." Detail from an Elizabethan broadside, showing women gossiping at the conduit while they wait in line to fill their buckets. To the right of the conduit, a quarrel is breaking out. The conduit is a typical medieval British fountain, with an enclosed tank and taps. The design of the fountain restricted access to the water and forced users to wait their turn. Reproduced in Christopher Hibbert, *London: The Biography of a City* (New York: William Morrow, 1969), 34.

brewed and steeped with conduit water were charged a yearly fee, but by 1497 such uses were again prohibited. Fishmongers, too, were forbidden to use the conduit. Professional water carriers in Paris were not permitted to draw water at the public fountains between sunset and sunrise, and then only if the basin was completely full. They were required to keep their yokes on their shoulders while waiting, so that when their turn came they could fill their buckets quickly, and they were banned from delivering fountain water to tradesmen such as dyers or horse dealers. In many cities, then, the main problem seems to have been that the supply of water available in public fountains was insufficient to meet the demands of both large-scale industrial users and the general public. Although civic officials tinkered with conduit policies in a futile attempt to satisfy all users, the domestic needs of the public were given priority.[20]

Modern studies indicate that when people have to wait in a long line for water from a public standpipe, they will revert to more readily available, albeit more risky, water sources. Many medieval Londoners, turning their backs on the overcrowded Great Conduit, continued to draw their water at the river. In 1324 Elena Gubbe drowned when she fell into the Thames from the stair of a wharf. She had gone down to the river at the hour of curfew with two earthenware pitchers for water. A nine-year-old girl named Mary met a similar fate in 1340, when she went down to the Thames one Sunday after vespers to fill an earthenware pot at a wharf. The cases of Elena and Mary may indicate a pattern of evening journeys to the river to draw water. Perhaps the water was to have been used in the preparation of the evening meal or for washing the kitchen equipment at the end of the day. Or the victims may have drowned because such evening journeys to the river were uncommon, and a person in distress was less likely to be rescued at a time when the waterfront was largely deserted. (Fatal accidents to bathers nearly always occurred in the evening, and the coroners' rolls sometimes specify that this was a time when no one else was around to witness the accident.) The social status of these unfortunate girls is unknown, but they may have been members of the "poor common people, who time out of mind have there [at the Thames bank] fetched and taken up their water." Londoners with the means to do so seem to have had their own wells, to have sent apprentices or servants to draw water, or to have purchased water from the city's water carriers.[21]

The degree to which London's professional water carriers transferred their water-collecting to the new water supply at the conduit remains an open question. The water-bearers do not seem to have been involved in any of the major disputes over the use of the conduit during the thirteenth and fourteenth centuries. The technology of the conduit was certainly designed to accommodate filling vessels such as tankards, but apparently not all members of the profession decided to utilize the new source. At least some water carriers continued to use the river as a water source long after the conduit was available—perhaps waiting to fill their tankards at the crowded conduit proved to be too time-consuming for vendors who depended on selling a large quantity of water each day. The conduit may have only recently come into operation when Henry Grene accidentally drowned while filling his tankard at the river in 1276, but water carriers were still utilizing the river in the fourteenth century. In 1325 John le Waterberere was on hand to raise the cry when a stabbing occurred on a wharf, and the river seems to have remained the source of supply for water-carts. Measures to clean up the dock at Dowgate were instituted in 1345 when it became so clogged with dung and other filth that the carters carrying water from the Thames "were no longer able to serve the commonalty." Weekly charges of 2d. for horses and 3d. for carts carrying water from the port were levied to keep the dock clean. The regulations on wages and prices issued in 1350 seem to show that water-carts were normally taking water collected from the river at Dowgate or Castle Baynard toward Chepe.[22]

Though the professional water carriers did not universally adopt the new hydraulic technology, neither did they universally reject it. Some water carriers were utilizing the conduit by the end of the fifteenth century. The guild regulations for water-bearers issued in 1496 specify that "no brother nor syster of the seid fraternyte shal have at the condyte at onys to his owne use above one tankard."[23] The regulation seems designed to facilitate access to the conduit: other guild members would not be forced to wait while one user filled multiple tankards; but the individual carrier's efficiency of scale, once he or she did obtain access to the conduit, was sacrificed. For the small-scale pedestrian water carrier who carried only one or two tankards, the conduit probably did provide a convenient supplement to the river as a water source; for the large-scale carrier employing a water-cart, its advantages were less apparent.

In London, then, only certain types of users transferred their activities to the conduit. With the exception of the fishmongers, who took advantage of the conduit's proximity to their market to wash their fish, the only groups that seem to have used the conduit were those who needed to draw water for domestic purposes or for trades such as brewing.[24] The conflicts between user groups at the conduit centered on two issues: supply and access. The structural configuration of the water system, with its lead supply pipes and a closed cistern, seems to have effectively protected the conduit from potentially polluting activities, such as washing clothes. The water was available as it flowed out from the spouts or taps, but the supply pipes and the cistern reservoir were not readily accessible. Since vessels were not lowered directly into the supply, their state of cleanliness was not an issue. Even when the fishmongers washed their fish at the conduit, objections were raised on the grounds that they were wasting water and impeding other users, not that they were polluting the supply. The disputes over the London conduit were essentially conflicts between competing social groups of users who employed the conduit for the same basic activity (drawing water), not between groups employing the conduit for functionally different purposes. Since the technological design was only really appropriate for drawing water, other patterns of water usage were not transferred to the conduit. This had the beneficial effect of eliminating certain types of disputes, but it also meant that regardless of the number of London's conduits, they could not replace the dependence on more traditional sources of supply for some categories of users.

Traditionally, horses and other animals were watered at rivers or ditches. In London this practice continued even after the Great Conduit was built. Italian fountains, with their open basins of water, were structurally better suited for watering horses, but the practice was considered undesirable, inasmuch as it compromised the quality of the domestic water supply. In order to prevent such inappropriate use of public fountains, municipal governments promulgated civic statutes and hired fountain wardens. In Viterbo, watering horses or other animals at Fontana Grande was specifically forbidden, and it is probable that such restrictions extended to all the city fountains.[25]

Legal restrictions alone proved insufficient, however. A better solution was found in the provision of a technological alternative, public watering troughs (*abbeveratoi*), to meet the needs of this group of users. By the middle of the thirteenth century, such facilities were being built in conjunction with Viterbo's new fountains, situated so that they could utilize the overflow water discharged from the fountain basin. This made good sense not only hydraulically but socially, and the provision of a convenient watering trough must have considerably reduced the temptation to use the fountains for this purpose. Like the fountains, Viterbo's troughs were regularly cleansed and were, in turn, legally protected against pollution and the diversion of their water.[26]

In Siena, as in Viterbo, the commune provided at the main public fountains specialized watering troughs, which were generally fed by the overflow from the main basin. In some cases they may have received independent supplies of water from veins that were considered subpotable.

FIG. 5.2. Horses and cattle drinking at a fountain in the Roman Forum. To protect the purity of the water in public fountains, some cities provided special watering troughs for animals. Drawing by Stefano della Bella (1636). Reproduced in Cesare d'Onofrio, *Le fontane di Roma* (Rome: Staderini Editore, 1957), fig. 85.

Bottino workmen were expected to differentiate between "acqua buona" and "acqua spugnosa." The former was channeled into the main fountain basins; the latter was thought to be suitable only for watering animals. As part of their policy of catering to the specialized needs of various users, the municipal government called for structural modifications to roads and piazzas near the fountains to ensure that animals had easy access to the watering troughs. The inappropriate use of the abbeveratoi was punished, pollution of the water being apparently the main concern. Siena even attempted to restrict the transmission of contagious diseases by infected animals: horses suffering from "capo morbo" were not allowed to use the public drinking troughs.[27]

The traditional practice of washing clothes at the riverbank was another activity that, if transferred to a public fountain, could result in the pollution of the water supply. To combat this problem, statutory prohibitions were frequently issued by municipal authorities. At Siena, for example, washing clothes in a main fountain or an abbeveratoio constituted an infraction and was penalized by a monetary fine. Washing clothes in the fountains was not an activity specifically prohibited by the Viterbese statutes, though it probably fell under the general antipollution ordinances. Such statutes were rare in England, probably because the design of most conduits did not easily lend itself to the practice. The wharf from which London's unfortunate Elena Gubbe had fallen was known as "La Lauenderebrigge," which suggests that it was a jetty used for washing clothes. An early-fifteenth-century ordinance forbidding the exclusion of common people from the wharves and stairs on the Thames bank specifically mentions beating and washing clothes as standard activities at the river. The fifteenth-century Leet Book of Coventry, however, did prohibit washing clothes at the conduit and assessed a 4d. fine for doing so.[28]

Both Siena and Viterbo reduced the temptation to use public fountains for washing clothes by providing users a convenient alternative, the specialized wash trough. Many Sienese fountains had a *lavatoio* (laundry trough) as a subsidiary basin, supplied with water from either the main basin or the abbeveratoio. The lavatoio itself was provided with stones for washing clothes. These were probably inclined slabs around the rim of the trough, which are familiar features in traditional European washhouses. In Viterbo the public lavatoi were fed by fountain water but were structurally separate entities. A lavatoio next to Fontana del Capone, which may date to the thirteenth century, is a long rectangular trough with sloping sides and a central channel fed by water coming from the fountain. Users are protected from the elements by walls that enclose three sides of the structure and support a roof. This general form is standard for Viterbo's postmedieval lavatoi and probably represents the basic medieval pattern. The long troughs permitted multiple users to wash their clothes simultaneously.[29]

Washing clothes was generally women's work. In Viterbo two statutes dealing with the "lavatorium de Rielli" assume that it is women who wash clothes there. Saint Bernardino of Siena acknowledged that washing clothes was normally a servant's task, but he felt that a devoted wife did a better job. Laundresses were found among the servants in royal, noble, and monastic establishments in England. Some washerwomen seem to have been free-lance professionals, although the profession was not well paid. Sienese tax assessments include washerwomen among the poorest 20 percent of taxpayers. Those who could afford it might prefer to employ servants or hire professional washerwomen for the task; nevertheless, the public wash-troughs, like the fountains, served an important social function by providing an opportunity for respectable women to spend time together outside the often restrictive privacy of their own homes.[30]

Men, at least occasionally, also washed clothes. The call for repairs to Siena's Fonte Nuova lavatoio in 1467 mentions both men and women as users. In England male launderers appear most often in association with monasteries, although even here the employment of washerwomen was more common. A washerman (*lotor*) is mentioned in the records of the bishop's visitations at some religious houses in the diocese of Lincoln, and a monk of Bardney Abbey was accused of committing adultery with the washerman's wife. The use of washerwomen by male religious houses could be a cause of grave concern. Huntington Priory was instructed to make sure that the women who washed the clothes waited at the outer gates rather than entering the inner precinct. The canons of Saint Frideswide's were told to either wash their garments themselves or to hire (male) fullers but not to employ women to do the job. At the Augustinian Priory at Barnwell, the laundress had to be a woman of good character: she was charged with mending and washing the community's surplices, sheets, shirts, and drawers once a fortnight in summer and once every three weeks in winter. A system of tallies was used to keep track of the items sent out to be laundered, and the washerwoman's wages were docked if any articles were missing. Monks may have washed some of their own clothes. Although Cluny sent the laundry out on Tuesdays, the brothers were permitted to wash their clothes out themselves in the cloister fountain, should they wish to do so. Unlike the nuns of Harrold Abbey, however, most men in religious orders would not have been reduced to washing their own clothes along the banks of the public river.[31]

Medieval water systems, unlike their Roman counterparts, were not designed primarily to accommodate bathers, but the use of conduits to supply baths was not entirely unknown. Monastic water systems sometimes supplied piped water to the bathhouse, and Westminster Palace had hot and cold running water piped to the royal bathtub. More commonly, however, bathing took place in rivers, in domestic bathtubs, or at specialized bathing establishments that remained independent of the public water systems. French pilgrims to Santiago de Compostela used to clean themselves in a wooded spot on the nearby river before paying their respects at the cathedral. For the love of the Apostle, they would wash "not merely their virile member, but having taken off their clothes, wash off the dirt from their entire body." Thirteenth-century Paris had twenty-six commercial establishments offering steam baths or tub baths.[32]

English bathhouses were generally supplied by water carriers. A complaint by the indentured Thomas Bunny charged that his mistress, who kept "stews" on the far side of London Bridge, had set him "all manner of grievous work, such as carrying water in tynes, and while thus employed he fell down and received a permanent injury." Citizens of Siena and Viterbo had recourse to bathing establishments at spas utilizing local hot springs. On occasion, however, irresponsible individuals attempted to use the fountains themselves for bathing. In 1282 Siena fined a certain Ciampolino three lire for an illicit bath that he had taken in Fonte de Follonico.[33]

To the extent that public fountains provided more convenient sources of domestic water, they may have had some impact on the frequency of home bathing and washing. Modern studies show that one of the benefits of providing public water systems to traditional societies is an increased incidence of washing when domestic water is more readily obtainable. The overall frequency of medieval washing and bathing is impossible to quantify (and doubtless showed considerable variation between social groups and individuals), but one gets the impression that washing the hands and face and even bathing the entire body became increasingly common activities in the later Middle Ages (albeit still infrequent by modern—or Roman—standards of personal hygiene). Frederick II's Sunday bath was considered scandalous by his contemporaries in northern Europe, whereas King John of England, whose household accounts indicate that he took ten baths within six months, was almost certainly more fastidious than most men of that time. The building accounts of John's successors, however, show that elaborate baths came to be considered standard features in royal palaces and manors during the later thirteenth and fourteenth centuries.[34]

Some monasteries were provided with specialized bathhouses, although how frequently they were used remains difficult to determine. The Canterbury plan shows a branch of the pipeline feeding water to a large bathhouse, which was staffed by attendants Cole, Milo, Richard, and Pagan. These large bathhouses would have contained a series of individual tubs. According to Lanfranc, monastic bathing days were to be conducted in a seemly and orderly manner, with the monks, screened by curtains, sitting in silence in the baths. Once he had washed himself, the monk was not to linger for pleasure but was to rise and get dressed promptly. Bathers at

FIG. 5.3. Although bathing in fountains was a popular artistic motif, as in this depiction of the legendary fountain of youth, citizens who tried to imitate art in their local civic fountains could be fined. Piedmontese Master, *The Fountain of Youth,* c. 1430. Reproduced in Liana Castelfranchi Vegas, *International Gothic Art in Italy* (London: Thames & Hudson, 1968), pl. 61.

Barnwell were given soap if they requested it; those at Abingdon were provided with hay on the floor.[35]

Monastic attitudes toward bathing reveal a tension between the demands of asceticism and those of ritual purity. Saint Benedict had discouraged frequent bathing for the young and healthy, and in many medieval monasteries, bathing was encouraged only before the great liturgical feasts, such as Christmas, Easter, and Pentecost. Saint Benedict himself, however, had prescribed baths for the sick, so it would seem that the demands of health could offer an opportunity for bathing at other times of the year. This loophole (analogous, perhaps, to the less restricted diet permitted those in ill health) makes it very difficult to estimate the actual frequency of monastic bathing. Even strict bishops like Eudes of Rouen, who were very concerned about other possible infractions of the rules, do not seem to have inquired closely into the bathing habits of the monks and nuns in their dioceses. Nonetheless, bathing that was seen as excessive could provoke criticism. The saintly Ailred of Rievaulx (who had built a little tank filled with icy water for penitential immersions) was accused by his detractors of giving up his body to baths and ointments. According to his biographer Walter Daniel, the criticism was unwarranted, since Ailred was tormented by stones in his urine and was forced to bathe frequently (up to forty times on one day) to soften the obstructions. Attitudes toward monastic bathing do seem to have become somewhat more relaxed by the end of the Middle Ages. By the fifteenth century even the Cistercians, who had strongly rebuked monks caught sneaking a bath outside the abbey in earlier years, were permitting healthy monks a monthly bath.[36]

Monks did wash their hands and faces on a daily basis, and it was customary to wash at set times, such as before entering the church and before and sometimes after meals (since they ate with a knife and their fingers). Mendicant friars, who spent more time outside the walls of the convent, presumably had a less regular schedule for washing: a Franciscan lavatory building mentioned in a Yorkshire court case was described as "a certain apartment where the friars commonly wash themselves when they come to the house tired and weary." If the house lacked a piped fountain-laver, water could be poured over the hands or directly into a basin. Salimbene tells the story of Brother Nicholas of Montefeltro, a friar who was so humble that when the dinner bell rang, he was always the first to come forward to pour water so that the other brothers could wash their hands.

Although somewhat inept, since he was corpulent and old, he performed this service with love and courtesy. Fountain-lavers were generally situated in the cloister, often near the door of the refectory, and sometimes enclosed in a fountain-house that extended into the cloister garth. Often they were provided with numerous jets or taps, so that multiple users could wash simultaneously. Towels would be kept hanging nearby, sometimes in a small cupboard. Saint Denis at Rheims had strict rules governing the proper use of the laver and towels: "If all cannot wash at the same time, the juniors are to wash first. . . . All the brethren are to be careful not to blow their noses with the towels, or to rub their teeth with them, or to stanch blood, or wipe off any dirt." Likewise, the authors of monastic customaries found it necessary to spell out prohibitions against practices that would sully the water, such as spitting, hawking phlegm, or blowing snot into the laver.[37]

Besides the daily washing of hands and faces, some monastic *lavatoria* were used for foot-washing, not only as part of the Maundy Thursday ritual but also as a pious practice at other times during the year. British Cistercians in particular seem to have designed their trough-lavers for dual-purpose washing. The monks stood in front of the trough when washing their hands, but for the weekly washing of the feet they sat on a special bench built above the basin, so that their feet could rest in the trough.[38]

Monastic lavatoria were also used for other activities. Commonly, monks would comb their hair either before or after washing at the laver. At Barnwell, sand and a whetstone were kept beside the laver so that the brothers could clean and sharpen their knives. Shaving of the beard and tonsure could also take place by the cloister laver or in a special shaving house, such as the one at Christ Church, Canterbury, which was located near the base of the lavatorium tower. In some establishments the brothers shaved each other, and if they were lucky, hot water would be provided. However, a razor in unskilled hands could prove to be a dangerous instrument. At Saint Augustine's Abbey in Canterbury, the brothers were sustaining numerous "injuries and sundry dangers . . . as they were without skill and knowledge in the work of shaving." Taking pity on his bloodied flock, Abbot Roger II instituted the practice of hiring a professional lay barber, who shaved the monks in a chamber next to the bathroom. Young monks seem to have had particular difficulties learning the art of shaving each other. At Bardney Abbey they were coming late to choir on shaving days, so that the high mass had to be celebrated without music. Their lack of skill was leading to such serious dangers that the abbey was ordered to hire a professional barber. (While their elders were being shaved, the young were to give their time to books and study, not waste their time on frivolities.)[39]

Secular equivalents of the monastic lavatoria were rare, though some wealthy Jewish households had courtyard fountains for ritual and hygienic hand-washing. Some individual washing probably took place by simply scrubbing the hands, face, or head in a basin of water. A more elaborate method was to have a servant hold a shallow basin under the cupped hands of the washer, while pouring water out of an ewer or an aquamanile. After washing, the hands were dried on a towel. For the late medieval nobility, quasi-ritualized washing at meals was a sign of refinement and breeding, and initiation into the protocols of personal hygiene was part of the instruction of noble children. It was not considered refined, for example, to spit in the basin or wipe one's nose on the towel. It is more difficult to determine how far down the social scale such concern with personal cleanliness extended. Late medieval literary references to bathing and artistic depictions of bathtubs are common. These motifs suggest that, for at least some classes of adults, bathing carried an aura of refinement, pleasure, conviviality, and amorous adventure, and bathing infants was seen as a sign of good parental care.[40]

Changing attitudes toward personal hygiene reflect broader cultural currents than the contemporary diffusion of hydraulic technology, but the two trends are, to some degree, intertwined. One the one hand, the increased demand for domestic water for washing and bathing would have stimulated the demand for public water systems; on the other, the provision of more convenient public water sources would have facilitated the diffusion of the new behavioral norms.

Industrial use of public water supplies was a perennial source of potential conflicts. As we have seen in the case of London's brewers, some tradesmen were accused of taking too much water away from domestic users. Certain other occupational practices threatened to pollute water supplies. Viterbo's leather workers, for example, were banned from using the water at Cripta Rielli, where the women washed their clothes. Siena barred leather and cloth workers from washing their products in the basins at Fonte Branda. Butchers, whose activities called forth countless municipal sanitation ordinances designed to protect streets and rivers from blood and entrails, also posed a threat to public fountains. In Coventry, entrails were not to be washed at the conduit. The Viterbo butchers' guild sought to police itself and strictly prohibited the cleaning or soaking of tripe, offal, heads, entrails, or anything else in the fountains.[41]

The cloth and leather industries in particular needed large quantities of water. Tradesmen in these industries often utilized river water (and used the river for wastewater disposal). According to a study of the distribution of tradesmen in York, based on the 1381 lay poll tax returns, all but three of the town's forty-four tanners lived in the parish of All Saints in North Street, which was situated on the west bank of the Ouse. Fullers were concentrated in the parish of Saint Mary and Margaret in Walmgate, near a branch of the river, but other textile workers were dispersed throughout the city. Urry's analysis of the Canterbury rentals shows several fullers and tanners holding ground near the river, although other members of the textile and leather industries were not concentrated in a particular locality.[42]

In some cities, textile and leather workers utilized more complex water systems. In medieval Winchester, first tanners and then fullers and cloth dyers occupied the houses and workshops that fronted onto Tanner Street (Lower Brook Street). Excavations there have revealed that the individual tenements were provided with private water channels leading off the artificial "brook," which ran down the center of the street. Flax and hemp retters in Viterbo had specialized pools (*piscine*) fed by artificial channels in the Piano di Bagni. Here they not only enjoyed an abundant water supply from the local hot springs, but the noxious odors emanating from the retting process were also well removed from the town.[43]

In Siena both the leather and cloth industries had their own piscine, which were regulated by the appropriate guilds. Like the specialized basins of the civic fountains, different industrial basins were reserved for different functions, according to the degree of pollution associated with the activity. The guildsmen seem to have emulated the civic fountains in designing the pools, and the civic authorities in their attempts to regulate water use by written statute. At least some of these pools were fed by the overflow from Fonte Branda. The industrial piscine were dependent on the public water system, and the Arte della Lana paid close attention to urban statutes that affected Branda's supply. Fonte Branda itself was provided with a fourth basin known as a *guazatoio*. There is no indication of the specific activity it accommodated. It may have been reserved for industrial functions, although its use for certain activities associated with leather and cloth processing was specifically prohibited. It seems to have been especially vulnerable to pollution, since a rigorous cleaning schedule for this particular basin was mandated in the civic statutes. The entire Fonte Vetrice was ceded by the city to the wool guild.[44]

The use of flushed channel drains supplemented but did not displace more traditional forms of waste disposal, such as rubbish pits, cesspits, or the direct dumping of wastes into streets or watercourses. Medieval latrines generally consisted of a seat made of a plank of wood with a hole in it, which was situated over a cesspit, stream, ditch, moat, or drain. Some cesspits would have been simple, unlined holes in the ground, but others could have different types of lining: excavated examples range from a barrel-lined shaft to elaborate stone-lined chambers. Sometimes a wooden pipe carried the wastes from the privy to a more remote cesspool. Latrines could be situated next to houses, in courtyards, or out in the backs of tenements. London building ordinances required that unlined cesspits be situated at least 3½ feet from the property line, whereas stone-lined pits could be 2½ feet from the neighbor's soil, but problems associated with latrines remained a frequent source of conflict between neighbors, to judge from the Assize of Nuisance complaints.[45]

Some cesspits were dangerously deep: Richard le Rakiere, seated on a latrine in his house, was drowned in a pool of sewage when the rotten planks suddenly gave way. Workmen digging a latrine pit in the courtyard of a London house and lining it with wine barrels had reached a depth of five casks when a board accidentally dropped out of one of barrels. When apprentice John de Aldinele climbed down a ladder to retrieve the board, he was overcome by bad air and the fumes from the casks and died of asphyxiation. The same fate befell his companion, John Putoys, who climbed down the shaft to see what was wrong. Channel drains may have been less hazardous, but even they could have their unexpected dangers. In 1184 the floor of the archbishop of Mainz's palace in Erfurt collapsed during a royal visit: a multitude of men fell into the great drain below and were flushed, along with the episcopal sewage, out to the river. King Henry VI and Archbishop Conrad, who had been sitting in recesses in the wall, were able to cling to columns in the windows and avoid a similar dunking.[46]

Monastic reredorters generally had a long row of cubicles, each with its own seat. Durham Priory's *necessarium* was remodeled in the early fifteenth century, so that it was "a most decent place" with wainscot partitions on either side of every seat "so that one of them could not see one another when they were in that place." At Canterbury, the night roundsman had the duty of taking a candle in a lantern and checking all the compartments in the reredorter. If a brother was found asleep on one of the seats, he was not to be touched but was to be awakened by a slight sound. Young monks who needed to visit the privy during the night were not to go unescorted: instead, they were to awaken their masters to accompany them and to light a lantern. The fear of solitary vice or homosexual encounters in the relative privacy of the latrine cubicles may underlie some of these monastic strictures, but the association of filth and demons was another source of anxiety. Salimbene tells the story of a young man in religious orders who was praising God while sitting on the privy. A demon accosted him and rebuked him for praying in such unsuitable circumstances, whereby the young man stoutly responded, "I shall praise God while emptying my bowels. For God abhors no filth except the filth of sin." He then went on to rout the demon, chastising him for skulking around the privy. "You were created to live in heaven, and now you seek out toilets and go visiting latrines." (Lest his readers develop apprehensions about visiting such demonic abodes from this tale, Salimbene reassures them that demons are easily confounded and put to shame.)[47]

Some medieval cities were provided with public toilets. London had at least thirteen, located along the city wall (where the effluent ran into the moat or the Walbrook stream) or by the Thames. Like monastic reredorters, some of these public "necessary houses" were very large. In 1306 spicer John le Spencer ordered a cask of wine to be delivered, then dodged the servant who was sent to fetch the money by telling him to wait for him while he went to the privy on London Bridge. While the trusting servant stood outside one entrance, Spencer made his escape out another door. In Exeter, a long, vaulted public toilet on the Exe Bridge was known as the Pixey or Fairy House. One of the most spectacular public privies was Whittington's Longhouse, built thanks to a charitable bequest by Richard Whittington, the famous mayor of London. The longhouse had two rows of sixty-four seats each, one side for women and one for men. The underlying gully, which emptied into the Thames, was flushed out into the river at high tide. Five almshouses for poor pensioners from the parish of Saint Martin Vintry topped off the imposing structure.[48]

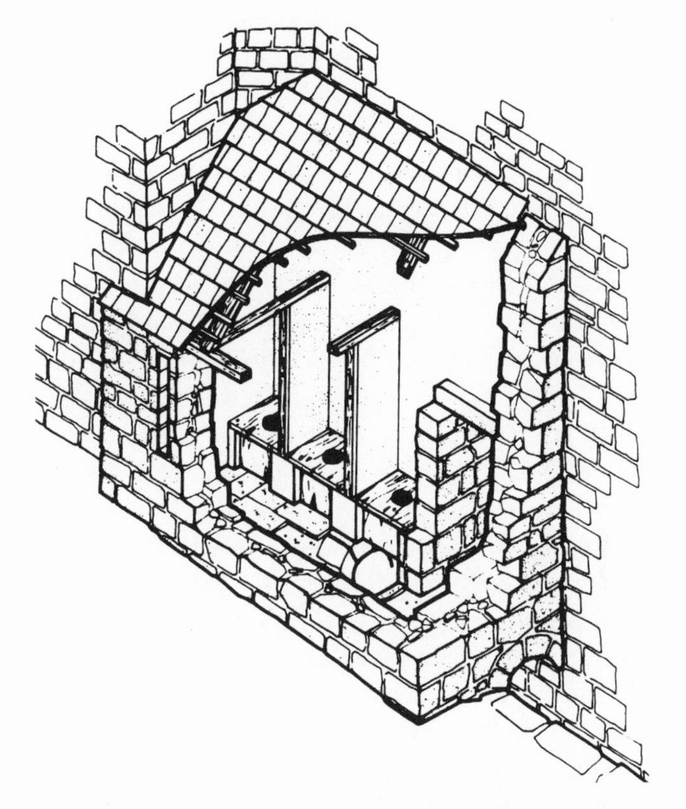

FIG. 5.4. Guest house latrines, situated over the main drain, Kirkstall Abbey. It was common to dump rubbish down medieval latrines, a practice that could block the drains. Stuart Wrathmell, *Kirkstall Abbey: The Guest House,* 2d ed. (Wakefield: West Yorkshire Archaeology Service, 1987), 2. Reproduced by permission of the West Yorkshire Archaeology Service.

In addition to using necessary houses for human wastes, medieval men and women often seem to have used them for the disposal of garbage of all sorts. A stone-lined latrine pit in Cuckoo Lane, Southampton, is thought to have belonged to the household of Richard of Southwick, a prominent burgess who died in 1290. It was filled with the debris of a prosperous household, including kitchen wastes, dead domestic animals (cats, dogs, sparrow hawks, a ferret, and a monkey), pottery, wooden bowls, old shoes, metal objects, rope, and baskets. The practice was harmless enough in the case of cesspits, but when the habit of indiscriminate dumping down the privy hole was transferred to latrines flushed by channel drains, problems could arise. As a result of these practices, channel sediments often contain substantial deposits of kitchen garbage (bone, shell, seeds), broken pottery, cloth, leather, and so forth. Ceramic vessels found in reredorter drains may have been used as urinals or chamber pots or may have held water for washing.[49]

In an analogous development, drain channels (some fitted with special disposal chutes) took the place of more traditional rubbish pits. At Bardney Abbey a stone sink set in the kitchen floor led into a wall outlet that fed into a drain. A groove for a wooden shutter was in the wall, and the sink may have been periodically flushed out with water stored in a tank above. The arrangement seems to have been designed for the disposal of kitchen refuse. Stone chutes leading into the main drain are visible in and near the kitchen at Furness Abbey. Such openings might be fitted with grates to screen out larger pieces; similar chutes also served as catch basins for runoff and other types of wastewater. Channel drains were not really suitable for this type of general waste disposal, since much of the material was too heavy for the flow to carry away—periodic manual cleaning was needed to remove the debris. At Christ Church, Canterbury, a workman cleaned out the main drain every Monday. Open watercourses, too, could become clogged by excess rubbish. London's city authorities authorized private latrines that emptied into the Walbrook, but only as long as the rubbish thrown in them did not impede the flow.[50]

Even cesspits required periodic emptying: the job was so unpleasant that latrine cleaners could command considerably higher wages than other unskilled workmen. Nevertheless, cultural attitudes toward excrement cast a social stigma on the occupation. Salimbene, who despised his townsman Gerard Segarello, expressed his opinion that Gerard was only suitable for vile occupations like cleaning latrines. (Gerard's offense had been to found a rival mendicant order known as the Apostles in their hometown of Parma, which garnered more alms from the townsmen than did Salimbene's own Friars Minor. If one is to believe Salimbene, Gerard had also taken Francis of Assisi's literal imitation of Christ to absurd extremes: among other excesses, he had himself circumcised, wrapped himself in swaddling clothes, lay in a cradle, and nursed at the breast of a young woman.)[51]

FIG. 5.5. Jugs used as a urinal and a stool pot. The ceramic vessels found in reredorter drains may have served similar purposes. *Voeux du Paon,* Pierpont Morgan Library, William S. Glazier collection, MS 24, fol. 27v. Franco-Flemish, mid–fourteenth century. Reproduced in Michael R. McCarthy and Catherine M. Brooks, *Medieval Pottery in Britain, AD 900–1600* (Leicester: Leicester Univ. Press, 1988), fig. 57.

The excavations at both Southampton and Worcester revealed numerous small pieces of cloth among the debris in the latrine pits. Such scraps of material may have served "to wipe the nether end," as did the cloths in Duke Humphrey of Gloucester's luxurious privy. (The ducal chamber also included cushions, hand-basins, and towels.) It is also possible that pieces of cloth would have served as sanitary napkins for medieval women. The Florentine Franciscan Brother Detsalve used the practice of wiping oneself with a piece of cloth to puncture the spiritual pretensions of the Dominican friar John of Vicenza. Having wrangled an invitation to lunch at John's convent, he reverently asked for a piece of John's tunic; then, retiring to the latrine, he relieved his bowels and wiped himself with the "relic," dropping it down into the sewage below. Detsalve was not content to keep the joke to himself. Shouting out, "Alas, help me brothers! I have lost the relic of a saint in the privy!" he lured the other friars to come help him find it. While each gullible friar bent over to peer down one of the seats, Detsalve enthusiastically stirred up the muck with a stick, "so that they might receive the full brunt of the stench," until even the pious Dominicans blushed in shame for having been fooled by such a notorious prankster.[52]

The successful adoption of complex intake and drainage systems necessitated the adoption of technologically appropriate behavior. The many statutes proscribing inappropriate practices document official attempts to enforce proper usage: penalties were usually monetary fines, though occasionally the threat of exile or prison was raised; in Siena a woman accused of deliberately poisoning the fountains was flayed alive and burned. (The Biccherna accounts for 1262 preserve a record of the expenditures arising from this gruesome execution.) Official laws could go only so far, however. Unofficial social sanctions must also have played a major role in discouraging inappropriate behavior, but unless the sanctions were ineffective and an incident escalated into a formal complaint or into violence, they were not likely to leave a trace in the written record. A riot that occurred in Viterbo in 1367 suggests that ordinary fountain users had their own standards of public hygiene and were prepared to verbally castigate those caught using a fountain improperly. According to the chronicles, certain members of the papal marshal's retinue were caught washing a puppy in one of the neighborhood fountains, the Fontana di Pianoscarano, whereupon they were scolded by one of the local women for polluting the drinking water. Tempers flared as the argument escalated, the woman was killed, and the entire neighborhood rose in a civic insurrection against the foreign members of the papal court in Viterbo. The ensuing riot was violently suppressed at the cost of many lives. To punish the unruly neighborhood, the fountain itself was demolished. Although this incident was unusual in its violent escalation and political implications, the threat of a public outcry among outraged neighbors must have effectively deterred many an abusive practice at the local fountain.[53]

The task of diffusing technologically appropriate behavior remained an ongoing challenge. Civic water systems were used by visitors and recent immigrants as well as by long-standing residents familiar with the fountains and city policies governing their use. Viterbo's legislators acknowledged that appropriate use of the city's fountains was something that had to be learned. Foreign visitors to Viterbo were not held liable for polluting a fountain or watering trough if they did so through ignorance of the city's statutes.[54]

A high proportion of the population of most medieval cities would have consisted of recent immigrants from the countryside. Case studies of medieval urban populations suggest that up to one-third or one-half of a city's inhabitants had come from rural villages. Patterns of water use and waste disposal that were acceptable practices in a small village could pose acute hazards in the densely occupied urban environment, and the new city dwellers had to learn to modify their traditional habits. Since a fountain, like the parish church, served as a focal point of a neighborhood community, many new urban immigrants probably first learned appropriate patterns of usage from members of their informal kin or friendship networks, who were already accustomed to urban ways. Because the rural influx into the city was a continuous flow, the problem of teaching new patterns of behavior was not confined to a single generation, nor could it be fully and finally resolved. The frequent practice of reissuing statutes related to water supplies and urban sanitation may not indicate the failure of such laws to modify the undesirable practices of a stubborn static population so much as the determination to meet the ongoing challenge posed by a flood of new urban immigrants.[55]

Some archaeologists suggest that the inhabitants of later medieval cities did enjoy a greater degree of urban hygiene than their early medieval predecessors, which may indicate that the efforts of civic legislators were at least partially successful. The willingness of neighbors to bring complaints to the attention of city courts also suggests that egregious polluters violated community norms and that redress was thought to be possible. Enforcing the sanitation ordinances must nevertheless have been, at times, a thankless task. In London William Bonet, the constable of Baynard Castle ward, was assaulted by an apprentice when he stopped him from emptying a handcart full of rubbish and filth into the Thames. Edmund le Coteler, relieving himself in the street, drew a knife on two men attempting to guard the streets of their ward from ordure; Beatrice Langbourne, upon being arrested for casting filth in the stree' accused alderman Simon de Worstede of being a "false thief and a broken-down old yokel."[56]

The adoption of complex hydraulic technology by users was not an unqualified success: not every activity requiring water was transferred to the new structures, nor was every activity that was transferred technologically appropriate. By and large, however, users seem to have accepted the new structures enthusiastically. This high level of acceptance can be attributed in part to the fact that the new technology did not place excessive new demands on its users. Fountains and conduits were not difficult to learn to use, and in most cases users were not expected to pay (at least directly) for the service. Nothing new had to be purchased to use the new water systems. The same water jars, tankards, and tynes that had served for river or well water were suitable vessels for drawing water at a conduit or a fountain. Equally importantly, using the new water systems did not violate customary motor patterns or traditional social interactions. Users were physically and socially comfortable with the technology. Complex water systems could be a financial burden and an administrative headache for municipal governments; but for urban residents, the benefits of the new systems far outweighed their drawbacks.

Epilogue

In July of 1538, Dr. John London issued an ominous report to Thomas Cromwell, concerning the state of the friaries in Oxford:

> *Grey Friars:* They have taken up the conduit pipes lately and cast them into sows to the number of 67, whereof 12 are sold, for the cost of taking up, as the warden says; the residue we have put in safeguard, and much of the conduit is not taken up.
>
> *Black Friars:* They have a very fair conduit.

The Oxford Greyfriars were only attempting to steal a march on the king's agents. The Protestant Reformation resulted in an abrupt disruption of monastic hydraulic technology throughout much of northern Europe. In Britain, water pipes were ripped out and melted down for their lead. Voids in the clay packings at Carmarthen retain the impressions of the Greyfriars' pipes, yanked from their trenches.[1]

Urban systems, too, could be the victims of political upheavals or sudden disasters. During a siege of Erfurt in 1309, the indignant defenders of the city were forced to watch helplessly from the walls as their conduit's extramural lead pipes were ripped out of their trenches, then hauled off and sold. Rebels in Exeter used lead from conduit pipes to make bullets during the Prayer Book Rebellion in 1549; during the Civil War, lead from the conduit and cistern in the Cathedral close was converted into "new warlike ammunitions." Lichfield Close had passed back and forth between the Royalists and the Parliamentarians, and a survey of 1652 found that the Cathedral's lead (including the conduit pipes, which had been dug up and cut off) had been embezzled and sold. All in all, England's Civil War seems to have been rather hard on lead components. At Worcester, 2,140 yards of pipes were removed from the Cathedral Priory conduit, along with the whole of the conduit house and the lead cisterns. Though London's water system survived the ravages of war, it was not so fortunate in the Great Fire of 1666. As the conflagration swept through Cheapside, it destroyed the public conduits in its path.[2]

The majority of medieval water systems, however, went out of use as the result of gradual decay and abandonment or were replaced with new waterworks in the postmedieval period. Surviving systems were likely to have their medieval components replaced during periodic overhauls. The pipes supplying the *lavatorium* at Maubuisson, for example, had apparently stopped flowing during the Hundred Years' War. The nuns tried to hire Paris fountain masters to fix the laver in 1489 but found the cost too high. A new laver fountain with new pipes was eventually installed in 1684.[3] Nonetheless, some medieval hydraulic structures still survive. Medieval lavatoria still stand in some monasteries, and medieval fountains still provide focal points of beauty and civic identity in European cities. Today their utilitarian functions have all but disappeared, but this is a recent development. Old photographs show Italian women still washing clothes in medieval *lavatoi* and still drawing water from medieval fountains, patterns of use that survived virtually unchanged from the thirteenth century into the twentieth.

Paradoxically, the very popularity of urban waterworks threatened their long-term survival. Since a primary goal of the systems was the conveyance of pure water, quality took precedence over quantity. In general, monastic water systems were able to satisfy the hydraulic requirements of their more restricted communities, but civic systems were under constant pressure to expand services. Successful public conduits created increased demands for additional fountains and for concessions for private branch pipes. The resulting overextension of distribution networks strained the capacities of systems to deliver water to consumers.

When cities were able to acquire new water sources and find adequate financing, they expanded existing systems or built new ones alongside them. In England, the suppression of the monasteries proved to be a hydraulic windfall for the towns. Bristol's monastic conduits were turned over to the parishes, which became responsible for their upkeep, and the Abbey conduit was still in everyday use in the nineteenth century. Henry VIII granted Cambridge's Franciscan conduit to Trinity College, where it still feeds the college fountain. In Coventry, Dr. London's report of 1538 indicated that the Greyfriars' conduit was much better than that of the town and that "much of the city shall lack water if they do not purchase it of the king." The mayors of Coventry followed his advice and modified and improved the conduit, which became part of the town water supply. In Lincoln, the city took over two friary conduits in 1539. The Blackfriars' system dated back to the thirteenth century, but the Greyfriars, unaware of the coming crisis, had completed their system just four years earlier. The city extended and modified the system, and it remained in use until the twentieth century.[4]

Not all communities were able to benefit from such chance windfalls, however. An alternative to system expansion was to scale back overextended distribution networks. Orvieto revoked private concessions, banned private fountains, suppressed some branches of its subterranean conduit, and eliminated several public fountains. In Paris, private pipes had become so prolific that a mere trickle of water was reaching the public fountains. In 1392 Charles VI revoked the grants relating to private branch pipes and ordered that the pipes be destroyed. (His zeal to restore the flow of water to the public was tempered, however, by self-interest and family feeling. He exempted the pipes belonging to himself, his uncles, and his brothers from the decree.) Although such measures temporarily increased the flow in the public fountains, they did not eliminate the underlying problem. The gap between urban supplies and consumer demands created a dynamic tension, which could escalate into an all-out tug-of-war between competing interests. Petitions for private diversions of public water supplies could be

FIG. E.1. High Street, Lincoln, c. 1835, showing the Tudor conduit still in use. The city was able to obtain the nearly new Greyfriars' water system at the Dissolution. Detail of a lithograph by I. Haghe. Reproduced in Andrew White, *St. Mary's Conduit, Lincoln,* Lincolnshire Museums Information Sheet, Archaeology Series, no. 19 (Lincoln: Lincolnshire County Council, 1980), 8.

difficult to refuse, particularly when they came from the powerful and wealthy. The improvements in the Paris system effected by Charles VI were undone by a rush for new private branch pipes in the sixteenth century, which was inaugurated when François I pressured the city authorities to permit his friend, the Bishop of Castres, to have a small tap ("the size of a pea only") in his house.[5]

Two main factors served as impediments to the unlimited expansion of traditional gravity-flow systems: the quantity of water available at the source and the high cost of conveyance. The pollution of rivers and urban groundwater meant that the goal of obtaining and distributing clean water was best met by tapping springs or deep aquifers. These water sources imposed inflexible topographical restraints, and the quantity of water they supplied was limited. Because conveyance systems were continuous structures, access to an unbroken conduit route between the source and destination was required, and money, influence, or political clout were needed to persuade local landowners to part with their land or to grant an easement. Masonry channels, lead pipes, and filtration conduits were also expensive. These "reverse salients" limited the capacity for growth, which led to a search for new responses to the problems imposed by resource limitations and costs.[6] Gravity-flow technology continued to be employed in the construction of some new water systems, but its dominance was increasingly challenged by the emergence of new hydraulic options.

The failure to find fully satisfactory solutions to the critical problems of quantity and cost set the stage for the emergence of new kinds of systems. Toward the end of the Middle Ages, certain key changes in hydraulic technology were implemented. The backers of the new systems responded to the clamor for more fountains and private pipes by redefining their priorities. Instead of providing a restricted quantity of pure water at a few outlets, they aimed for the widespread distribution of a plentiful supply of lower-quality water. With the shift in emphasis from water quality to water quantity, designers of the new systems turned to urban rivers as their source, which freed them from the topographic restrictions and quantitative limitations inherent in a dependence on springs. Artificial lifting devices—waterwheels, pumps, and engines that combined the two—sidestepped the old problems of long-distance conduit routes, topographic gradients, and hostile landowners. As long as a town was situated on a dependable river, water could be raised at any convenient location, stored in a water tower, and distributed directly throughout the town by means of pipes. The higher and continuous water pressures in the new systems helped ensure longer life spans for wood pipes. Since wood pipes were more economical than lead or earthenware conduits, and since long-distance pipelines were no longer necessary, conveyance costs were reduced. The postmedieval period also witnessed the construction of some ambitious channel-intake systems, supplied by long, river-fed canals, such as London's New River waterworks. Many towns that had never had complex water systems in the Middle Ages acquired them in the early modern period; towns that already had medieval water systems replaced or supplemented them with systems based on the new technology clusters. The old water systems did not completely disappear, but from the end of the Middle Ages they had to compete with what must have appeared to be very attractive new technological options: hydraulic systems that could deliver seemingly limitless quantities of water directly to individual houses.[7]

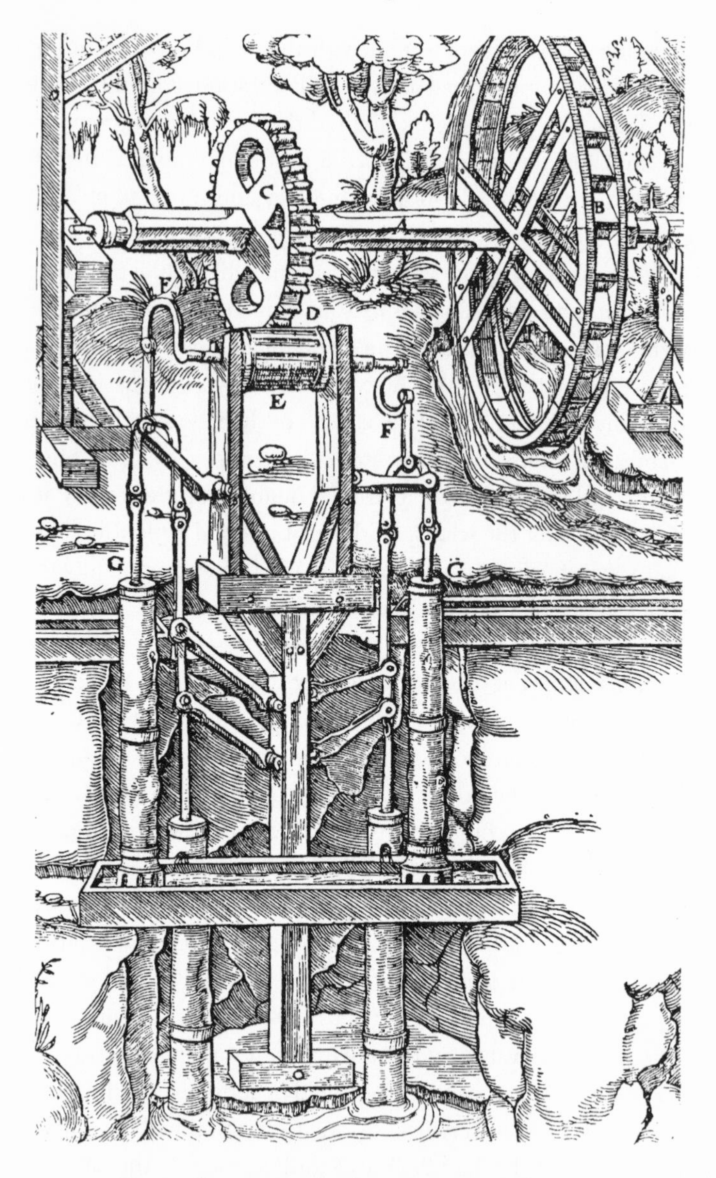

FIG. E.2. Design for a engine for draining mine shafts. Complex water engines composed of pumps and waterwheels were also used to raise water for urban water systems. Woodcut from 1556 edition of Agricola's *De Re Metallica.* Georgius Agricola, *De Re Metallica,* trans. Herbert Clark Hoover and Lou Henry Hoover (New York: Dover Publications, 1950), 189.

The individual technological components of artificial lifting devices were not new. Water-lifting wheels and pumps had been known in antiquity, and norias were widespread in the Islamic world (including Spain). There is some indication that fairly elaborate lifting devices had occasionally been used to raise water from wells in medieval Europe, but they do not seem to have been used in conjunction with complex conveyance systems until fairly late. By the late thirteenth century, water-lifting wheels were feeding complex systems in a few north German towns. Pumps were added to create more complex water engines as the practice spread south through Germany to Switzerland in the fourteenth and fifteenth centuries. Since similar engines were being developed to drain water from mine shafts, it is probable that there were close links between advances in mining technology and the new urban water systems.[8]

Hydraulic devices had a prominent role in the popular Renaissance "machine books" (*theatrum mechanorum*), which illustrated the latest technological advances and dazzled readers with drawings of imaginative mechanical novelties extrapolated from existing technologies. A fascination with ingenious hydraulic mechanisms (including lifting wheels and pumps) plays a significant part in the mid-fifteenth-century engineering drawings of Siena's Mariano Taccola, and later works exhibit a similar enthusiasm. Printed illustrations and their accompanying commentaries seem to have inspired some of the playful, theatrical waterworks found in Renaissance gardens.[9]

Pumps began to appear in England in the late fifteenth century. In 1581 Peter Morris (a Dutch or Flemish man) built a water engine within the first arch of London Bridge. The waterwheel worked force pumps, which supplied Thames water directly to individual houses in the eastern part of the city by means of lead and wood pipes. This was a new departure for London: instead of the old public conduits, Morris's system was privately owned and operated, with customers paying fees for the service. Private ventures in supplying water systems became more common in the seventeenth century, as shareholders obtained patents and invested in profit-seeking schemes to lift and convey water. Derby engineer George Sorocold built a new, improved water engine at London Bridge in 1701 and was responsible for the design and installation of water engines in many provincial English towns in the late seventeenth and early eighteenth centuries.[10]

The popular new systems provided more copious and convenient sup-

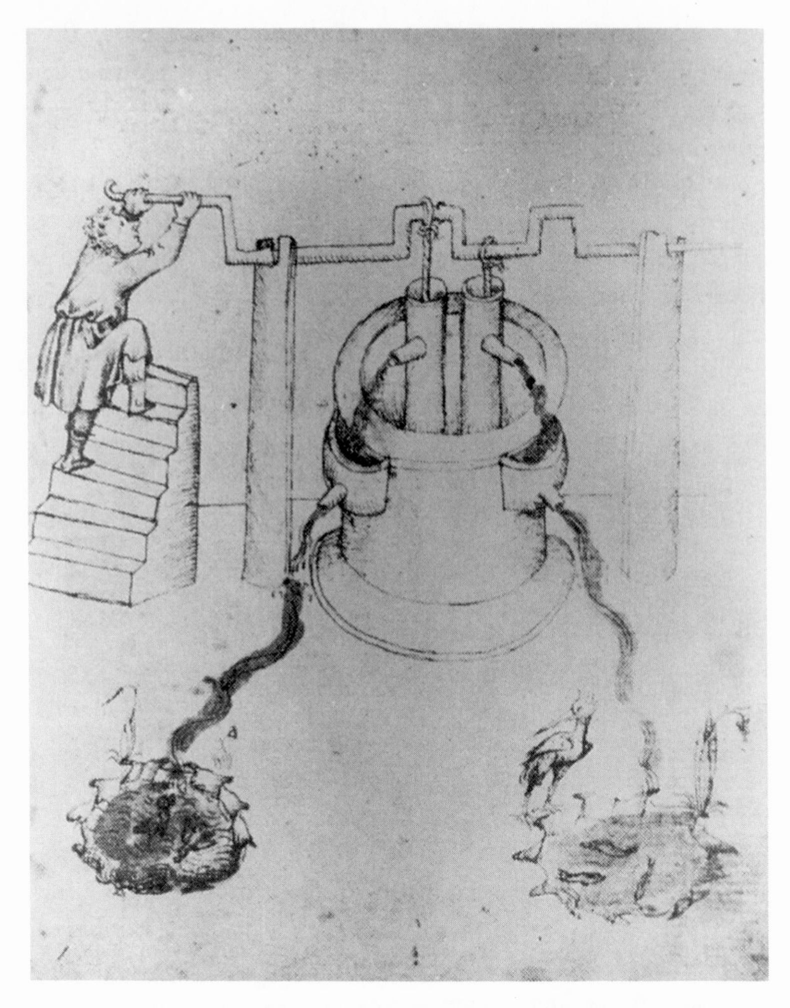

FIG. E.3. Drawing of a piston pump by Mariano Taccola. Drawings of clever hydraulic devices were characteristic features of Renaissance "machine books," and the popularity of such works helped spread technological awareness. Taccola came from Siena, a city with a rich tradition of hydraulic engineering. Mariano Taccola, *De Ingeneis,* Palat 766, bk. 3, fol. 29r. Reproduced in Frank D. Prager and Gustina Scaglia, *Mariano Taccola and His Book De Ingeneis* (Cambridge: MIT Press, 1972), fig. 25.

FIG. E.4. George Sorocold's water engine at London Bridge replaced the old Morris waterworks. Sorocold installed water engines in many English towns. Walter Minchinton, *Life to the City: An Illustrated History of Exeter's Water Supply from the Romans to the Present Day* (Newton Abbot, Eng.: Devon Books, 1987), 22.

plies of water than their medieval predecessors, but they came with hidden costs. The medieval social ethic, which had sought to guarantee water to all classes and which defended the interests of the poor, was threatened. Owners and shareholders of private water companies were only concerned with supplying paying customers. Where public conduits remained, the poor could still use them, but maintaining a public service became a lower priority for well-to-do citizens once private water services were installed. In London and elsewhere, large medieval conduits located in the streets were becoming serious traffic obstructions. Both the public Great Conduit system and Morris's private water engine were destroyed in the Great Fire of 1666. Although there was talk of restoring the old public conduits, nothing was done, whereas Morris's engine was quickly rebuilt. Some new systems were sponsored by municipal councils, but private companies played an increasingly important role in the supply of urban water, so that by 1846, only 10 out of some 190 British local authorities controlled their city's waterworks.[11]

The use of untreated river water to supply the new systems placed consumers at an increased risk of exposure to waterborne diseases, a problem that became more acute with the rapid growth of urban populations and the intensification of industrial pollution during the Industrial Revolution. Though customers were probably generally aware that the quality of water delivered by the new systems left something to be desired, they were unable to perceive the full extent of the health risk; the benefits that the new systems offered, however, were readily apparent.

The public health implications of the new technology did not become fully apparent until the mid-nineteenth century, when the role of contaminated water in the transmission of disease was first scientifically demonstrated. In England, the cholera epidemics of 1848–49 and 1853–54 triggered a series of parliamentary reports, which addressed the issues of sanitation and water purity and called for long-distance, municipally sponsored conduits to be supplied by pure, external water sources. The Board of Health report, issued in 1850, decried the quality of water being distributed by London's private water companies. It called for a new supply of pure water to be consolidated under public management and turned to history to back up its argument. "It has from earliest times been recognized as the duty of Government to take cognizance of running waters." In support of this claim, several medieval documents pertaining to monastic conduit licenses and river sanitation were enumerated. The report's authors, however, seem to have been unaware that their very own British municipalities had been providing just such a public service only a few centuries before.[12]

The Royal Commission on Water Supply's report, issued in 1869, again invoked the past. In their call for municipal waterworks, the commissioners recognized that London had, "at a very early date" been supplied by spring-fed conduits. The report continues: "The supercession of the municipalities by joint-stock companies is a comparatively modern innovation . . . of late years many towns in England have come to the conclusion that the new practice was a fundamental error, and have resumed the ancient principle by taking the control of the water supply again into their own hands. . . . A sufficiency of water supply is too important a matter to all classes of the community to be made dependent on the profits of an association. We are hence led to the conclusion that future legislation should restore the ancient practice."[13] The Victorian reformers, with their call for municipally sponsored water systems, their concern with the provision of water to all classes, and their emphasis on water quality, were not just being romantically anachronistic in their appeal for a return to an "ancient" practice. Their knowledge of hydraulic history may have been somewhat imprecise, but they managed to invoke quite correctly the key features of medieval water systems as they raised the cry for a new hydraulic revolution.

Abbreviations

Ant.J.	*The Antiquaries Journal*
Arch.Cant.	*Archaeologia Cantiana*
Arch.J.	*The Archaeological Journal*
BAR	British Archaeological Reports
B-P	Fabio Bargagli-Petrucci. *Le Fonti di Siena e i loro aquedotti.* 2 vols. Siena: L. S. Olschki, 1903.
CA	*Canterbury's Archaeology*
Cal.Chart.R.	Great Britain. Public Record Office. *Calendar of the Charter Rolls Preserved in the Public Record Office.* London: HMSO, 1903–.
Cal.Cl.R.	Great Britain. Public Record Office. *Calendar of the Close Rolls Preserved in the Public Record Office.* London: HMSO, 1892–.
Cal.Cor.R.	Reginald R. Sharpe, ed. *Calendar of Coroners' Rolls of the City of London AD 1300–1378.* London: Clay, 1913.
Cal.Dublin	John T. Gilbert, ed. *Calendar of Ancient Records of Dublin.* Vol. 1. Dublin: J. Dollard, 1889.
Cal.E.M.R.	A. H. Thomas, ed. *Calendar of Early Mayor's Court Rolls Preserved among the Archives of the Corporation of the City of London at the*

	Guildhall, AD 1298–1307. Cambridge: Cambridge Univ. Press, 1924.
Cal.Glouc.	W. H. Stevenson, ed. *Calendar of the Records of the Corporation of Gloucester.* Gloucester: Bellows, 1893.
Cal.L-B	Reginald R. Sharpe, ed. *Calendar of Letter-Books Preserved among the Archives of the Corporation of the City of London at the Guildhall.* London: Francis, 1899–1912.
Cal.P&M.L.	A. H. Thomas, ed. *Calendar of Plea and Memoranda Rolls Preserved among the Archives of the Corporation of the City of London at the Guildhall.* Cambridge: Cambridge Univ. Press, 1926.
Cal.Pat.R.	*Calendar of the Patent Rolls Preserved in the Public Record Office.* London: HMSO.
Cal.Wells	Great Britain. Historical Manuscripts Commission. *Calendar of the Manuscripts of the Dean and Chapter of Wells.* Vol. 1. London: HMSO, 1907.
Cal.Wills	Reginald R. Sharpe, ed. *Calendar of Wills Proved and Enrolled in the Court of Husting, London AD 1258–AD 1688.* 2 vols. London: Francis, 1889.
CBA	Council for British Archaeology
Con.1262	Siena (Italy). *Il constituto del comune di Siena dell'anno 1262.* Ed. Lodovico Zdekauer. Milan: U. Hoepli, 1897.
Con.1309	Siena (Italy). *Il costituto del comune di Siena volgarizzato nel MCCCIX–MCCCX.* Siena: L. Lazzeri, 1903.
EETS	Early English Text Society
HKW	H. M. Colvin et al. *The History of the King's Works.* 6 vols. London: HMSO, 1963–82.
JAS	*Journal of Archaeological Science*
JBAA	*Journal of the British Archaeological Association*
L&P H VIII	Great Britain. Public Record Office. *Letters and Papers, Foreign and Domestic, of the Reign of Henry VIII.* Ed. J. S. Brewer. 2d ed. Vaduz, [Lichtenstein]: Kraus Reprint, 1965.
LA	*The London Archaeologist*
LTR	*London Topographical Record*
LMPME	Skelton, R. A., and P. D. A. Harvey. *Local Maps and Plans from Medieval England.* Oxford: Clarendon Press, 1986.
MA	*Medieval Archaeology*
MEFRM	*Mélanges de l'Ecole française de Rome Moyen Age*
MGH	Monumenta Germaniae historica
PBSR	*Papers of the British School at Rome*
PL	Patrologiae cursus completus . . . series latina. Ed. J.-P. Migne.
PSANHS	*Proceedings of the Somerset Archaeological and Natural History Society*

PSIAH	*Proceedings of the Suffolk Institute of Archaeology and History*
RCHM	Great Britain. Historical Manuscripts Commission. *Report of the Royal Commission on Historical Manuscripts*. London: HMSO, 1870–.
RS	Rerum britannicarum medii aevi scriptores (Rolls Series)
SS	Selden Society
Stat.1237–38	Statuto di Viterbo del 1237–38. P. Egidi, ed., "Gli statuti viterbese del MCCXXXVII–VIII, MCCLI–II e MCCCLVI." In *Statuti della Provincia Romana,* ed. V. Federici, R. Morghen, P. Egidi, A. Diviziani, O. Montenovesi, F. Tomassetti, and P. Fontana. Fonti per la Storia d'Italia 69. Rome: Tipografia del Senato, 1930. 29–282.
Stat.1251–52	Statuto di Viterbo del 1251–52. P. Egidi, ed., "Gli statuti viterbese del MCCXXXVII–VIII, MCCLI–II e MCCCLVI." In *Statuti della Provincia Romana,* ed. V. Federici, R. Morghen, P. Egidi, A. Diviziani, O. Montenovesi, F. Tomassetti, and P. Fontana. Fonti per la Storia d'Italia 69. Rome: Tipografia del Senato, 1930. 29–282.
TBGAS	*Transactions of the Bristol and Gloucestershire Archaeological Society*
TLMAS	*Transactions of the London and Middlesex Archaeological Society*
TSSAHS	*Transactions of the South Staffordshire Archaeological and Historical Society*
VCH	The Victoria History of the Counties of England.
WIM	Frontinus-Gesellschaft. *Die Wasserversorgung im Mittelalter.* Geschichte der Wasserversorgung, no. 4. Mainz am Rhein: P. Von Zabern, 1991.
WWHHC	Richard G. Feachem, Michael McGarry, and Duncan Mara, eds. *Water, Wastes, and Health in Hot Climates*. London: Wiley, 1977.

Notes

PREFACE

1. *Cal.Cor.R.*, 219–220.

2. Bernard of Clairvaux, "Sermo de aquaeductu. In nativitate beatae Mariae virginis," PL, vol. 183, cols. 440–442; "Sancti Bernardi abbatis Clarae-Vallensis vita et res gestae. Libri septem comprehensae," PL, vol. 185, col. 285.

3. Joseph Thomas Fowler, ed., *Rites of Durham,* Surtees Society, no. 107 (Durham: Andrews & Co., 1903), 82–83; Joseph Thomas Fowler, ed., *Extracts from the Account Rolls of the Abbey of Durham,* 3 vols., Surtees Society, nos. 99, 100, 103 (Durham: Andrews & Co., 1898–1901). See, for example, 2:525; W. H. St. John Hope and J. T. Fowler, "Recent Discoveries in the Cloister of Durham Abbey," *Archaeologia* 58, no. 2 (1903): 437–460.

4. Thomas Hughes, in his study of electricity networks, almost invites a comparative analysis of modern and medieval technological systems: "In a culture that did not calculate capital costs—the medieval Western civilization for instance—electric light and power systems would have grown differently." As the Middle

Ages drew to a close and speculators began to build new private water systems based on capitalist calculations, European water systems did indeed grow differently. Thomas P. Hughes, *Networks of Power: Electrification in Western Society, 1880–1930* (Baltimore: Johns Hopkins Univ. Press, 1983), 463.

5. See Everett M. Rogers, *Diffusion of Innovations,* 4th ed. (New York: Free Press, 1995); Hughes, *Networks of Power;* Wiebe E. Bijker, Thomas P. Hughes, and Trevor J. Pinch, eds., *The Social Construction of Technological Systems: New Directions in the Sociology and History of Technology* (Cambridge: MIT Press, 1987).

CHAPTER 1. SURVIVAL AND REVIVAL

1. A similar demonstration, using a pipe eighteen feet high, was conducted as the pipe approached the close. *LMPME,* 67–69.

2. Agnolo di Tura del Grasso, "Cronaca senese," in *Cronache Senesi,* ed. Alessandro Lisini and Fabio Iacometti, Rerum Italicarum scriptores, new ed., vol. 15, no. 6 (Bologna: N. Zanichelli, 1931–39), 537; Anne Coffin Hanson, *Jacopo della Quercia's Fonte Gaia* (Oxford: Clarendon Press, 1965), 7–8, 106.

3. Bryan Ward-Perkins, *From Classical Antiquity to the Middle Ages: Urban Public Building in Northern and Central Italy,* A.D. 300–850 (Oxford: Oxford Univ. Press, 1984), chap. 7 and app. 3; Carlo Troya, ed., *Codice Diplomatico Longobardo dal DLXVIII al DCCLXXIV,* Storia d'Italia del Medio-Evo, no. 4.5 (Naples: Stamperia Reale, 1855), vol. 5, nos. 759–762; Bertram Colgrave, ed. and trans., *Two Lives of St. Cuthbert* (New York: Greenwood Press, 1969), 242–245; Liudprand, Bishop of Cremona, "Opera," in *Quellen zur Geschichte der sächsischen Kaiserzeit: Widukinds Sachsengeschichte, Adalberts Fortsetzung der Chronik Reginos, Liudprands Werke,* comp. Albert Bauer and Reinhold Rau (Darmstadt: Wissenschaftliche Buchgesellschaft, 1971), 282, 360; Anonymous ticinensis [Opicino de Canistris], *Liber de laudibus civitatis ticinensis,* ed. Rodolfo Maiocchi and Ferruccio Quintavalle, Rerum Italicarum scriptores, new ed., vol. 11, pt. 1 (Città di Castello: S. Lapi, 1903), 20.

4. André Guillerme, "La destruction des aqueducs romains des villes du nord de la France," in *Journeés d'études sur les aqueducs romains,* ed. J.-P. Boucher (Paris: Société d'Edition "Les Belles Lettres," 1983), 167–173; Guilhem Fabre, *The Pont du Gard: Water and the Roman Town,* trans. Janice Abbot (Paris: Presses du CNRS, 1992), 98.

5. Roberta J. Magnusson, *Medieval Water Supplies: Hydraulic Technology and Social Organization in England and Italy* (Ph.D. diss., Univ. of California, Berkeley; Ann Arbor, Mich.: University Microfilms, 1994), chap. 1 nn. 57–58.

6. Michael Greenhalgh, *The Survival of Roman Antiquities in the Middle Ages* (London: Duckworth, 1989), 109–110; Guillerme, "La destruction des aqueducs";

Klaus Grewe, "Römische Wasserleitungen nördlich der Alpen," in *Die Wasserversorgung Antiker Städte,* Geschichte der Wasserversorgung, no. 3 (Mainz am Rhein: P. von Zabern, 1988), 2:93–94.

7. Walter Horn and Ernest Born, *The Plan of St. Gall: A Study of the Architecture and Economy of, and Life in a Paradigmatic Carolingian Monastery* (Berkeley: Univ. of California Press, 1979), 1:69; W. C. Wijntjes, "The Water Supply of the Medieval Town," *Rotterdam Papers* 4 (1982): 199; Ward-Perkins, *From Classical Antiquity,* 189; L. A. Muratori, ed., *Antiquitates Italicae Medii Aevi* (Milan: Societatis Palatinae, 1738–42), 2:453D–455A; Clemens Kosch, "Wasserbaueinrichtungen in hochmittelalterlichen Konventanlagen Mitteleuropas," in *WIM,* 92; Joseph Stevenson, ed., *Chronicon Monasterii de Abingdon,* RS, no. 2 (London: Longman, Brown, Green, Longmans & Roberts, 1858), 2:270, 278.

8. Paul Benoit and Monique Wabont, "Mittelalterliche Wasserversorgung in Frankreich. Eine Fallstudie: Die Zisterzienser," in *WIM,* 191–192; Charles B. McClendon, *The Imperial Abbey of Farfa: Architectural Currents of the Early Middle Ages,* Yale Publications in the History of Art, no. 36 (New Haven, Conn.: Yale Univ. Press, 1987), 6, 8; Otto Lehmann-Brockhaus, *Schriftquellen zur Kunstgeschichte des 11. und 12. Jahrhunderts für Deutschland, Lothringen und Italien,* 2 vols. (Berlin: Deutscher Verien für Kunstwissenschaft, 1938), 1:90; Paolo Squatriti, *Water and Society in Early Medieval Italy,* A.D. *400–1000* (Cambridge: Cambridge Univ. Press, 1998), 17–18.

9. Lehmann-Brockhaus, *Schriftquellen zur Kunstgeschichte,* 1:31, 80, 82, 98, 128, 372, 379–380, 392, 413; *WIM,* 44, 55, 91–94, 104, 122–123, 237–257, 268–271.

10. André Guillerme, "Puits, aqueducs et fontaines: L'alimentation en eau dans les villes du nord de la France, Xe–XIIIe siècles," in *L'eau au moyen âge,* Senefiance, no. 15 (Aix-en-Provence, Marseille: Publications du CUER MA, Univ. de Provence, 1985), 185–200; Heinz Dopsch, "Der Salzburger Almkanal," in *WIM,* 282–286; "De Gallica profectione Domni Petri Damiani et ejus ultramontano itinere," PL, vol. 145, cols. 873–874; Orderic Vitalis, *The Ecclesiastical History of Orderic Vitalis,* ed. and trans. Marjorie Chibnall (Oxford: Clarendon Press, 1978), 6:208–209; "Sancti Bernardi abbatis Clarae-Vallensis vita et res gestae. Libri septem comprehensae," PL, vol. 185, cols. 285, 569–574; Pietrantonio Pace, *Gli acquedotti di Roma e il De Aquaeductu di Frontino,* 2d ed. (Rome: Art Studio S. Eligio, 1986), 193; Lehmann-Brockhaus, *Schriftquellen zur Kunstgeschichte,* 1:491–492, 508; Maria Stella Calò Mariani, "Utilità e diletto. L'acqua e le residenze regie dell'Italia meridionale fra XII e XIII secolo," *MEFRM* 104, no. 2 (1992): 343–372; Martin Biddle, "Wolvesey: The *domus quasi palatium* of Henry de Blois in Winchester," in *Chateau Gaillard: European Castle Studies,* ed. A. J. Taylor (London: Phillimore, 1969), 3:33; Otto Lehmann-Brockhaus, *Lateinische Schriftquellen zur Kunst in England, Wales und Schottland vom Jahre 901 bis zum Jahre 1307,* 5 vols.

(Munich: Prestel, 1955–60), 1:441; 2:602–603; C. James Bond, "Water Management in the Rural Monastery," in *The Archaeology of Rural Monasteries,* ed. Roberta Gilchrist and Harold Mytum, BAR British Series, no. 203 (Oxford: BAR, 1989), 83–111; C. James Bond, "Water Management in the Urban Monastery," in *Advances in Monastic Archaeology,* ed. Roberta Gilchrist and Harold Mytum, BAR British Series, no. 227 (Oxford: BAR, 1993), 43–78; *HKW,* 1:549; B-P, 2:68; Nevio Basezzi and Bruno Signorelli, *Gli antichi acquedotti di Bergamo* (Bergamo: Comune di Bergamo, 1992), 60; Cesare Pinzi, *Storia della città di Viterbo,* 4 vols., Biblioteca Istorica della Antica e Nuova Italia, no. 195 (Rome: Tip. della Camera dei Deputati, 1887–99), 1:242–243.

11. Anna Götlind, "Technology and Religion in Medieval Sweden" (Ph.D. diss., Univ. of Göteborg, Sweden, 1993), 51–56.

12. Conrad of Eberbach, quoted in C. H. Lawrence, *Medieval Monasticism: Forms of Religious Life in Western Europe in the Middle Ages,* 2d ed. (London: Longman, 1989), 182. M.-D. Chenu, *Nature, Man, and Society in the Twelfth Century: Essays on New Theological Perspectives in the Latin West,* trans. Jerome Taylor and Lester K. Little (Chicago: Univ. of Chicago Press, 1968), 4–18, 37–48.

13. Magnusson, *Medieval Water Supplies,* chap. 3.

14. Everett M. Rogers, *Diffusion of Innovations,* 4th ed. (New York: Free Press, 1995), 286–290.

15. Duccio Balestracci, "Siena e le sue fonti," in *Siena e l'acqua: Storia e immagini di una città e delle sue fonti,* ed. Vinicio Serino (Siena: Nuova Immagine Editrice, 1997); Richard Krautheimer, Spencer Corbett, and Alfred K. Frazer, *Corpus Basilicarum Christianarum Romae (IV–IX Cent.)* (Vatican City: Pontifico Istituto di Archeologia Cristiana, 1977), 5:173–176, 229–230, 262–264, 267, 271; Francis Morgan Nichols, ed. and trans., *Mirabilia urbis Romae: The Marvels of Rome* (London: Ellis and Elvey, Spithoever, 1889), 73–74; Agostino Paravicini Bagliani, "La mobilità della curia romana nel secolo XIII, riflessi locali," in *Società e istituzioni dell'Italia comunale: L'esempio di Perugia (Secoli XII–XIV)* (Perugia: Dep. Storia Patria, 1988), 155–278; Laure Beaumont-Maillet, *L'eau à Paris* (Paris: Hazan, 1991), chap. 3; F. W. Robins, *The Story of Water Supply* (London: Oxford Univ. Press, 1946), 136–137; William Melczer, *The Pilgrim's Guide to Santiago de Compostela* (New York: Italica Press, 1993), 122; Luis Vázquez de Parga, José María Lacarra, and Juan Uría Ríu, *Las peregrinaciones a Santiago de Compostela* (Madrid: n.p., 1948), 1:56–57; B. A. Windeatt, trans., *The Book of Margery Kempe* (Harmondsworth: Penguin, 1985), 147–148. Klaus Grewe, "Der Wasserversorgungsplan des Klosters Christchurch in Canterbury (12 Jahrhundert)," in *WIM,* 235; Stuart Wrathmell, *Kirkstall Abbey: The Guest House,* 2d ed. (Wakefield: West Yorkshire Archaeology Service, 1987), 12–15; William, Archbishop of Tyre, *A History of Deeds Done beyond the Sea,* trans. Emily Atwater Babcock and A. C. Krey (New York: Columbia Univ. Press, 1943), 202–203; Yehuda Peleg, "The Water Supply System of

Caesarea," *Leichtweiss-Institut für Wasserbau der Technischen Universität Braunschweig: Mitteilungen* 82 (1984): 3; Knut Olof Dalman, *Der Valens-Aquädukt in Konstantinopel,* Istanbuler Forschungen, no. 3 (Bamberg, Germany: J. M. Reindl, 1933), 8–10.

16. Everett M. Rogers, *Diffusion of Innovations,* 3d ed. (New York: Free Press, 1983), 170; Biddle, "Wolvesey," 31; Giraldus Cambrensis, *Opera,* ed. James F. Dimock, RS, no. 21g (London: Longmans, Green, 1877), 7:45; Jim Gould, "The Twelfth-Century Water Supply to Lichfield Close," *Ant.J.* 56 (1976): 75; Clemens Kosch, "Die Wasserleitung vom Ende des 11. Jahrhunderts in ehem. Kloster Grosskomburg," in *WIM,* 237–243; Valentine Jackson, "The Inception of the Dodder Watersupply," in *Medieval Dublin: The Making of a Metropolis,* ed. Howard Clarke (Dublin: Irish Academic Press, 1990), 128–141; M. D. Lobel and E. M. Carus-Wilson, "Bristol," in *The Atlas of Historic Towns,* ed. M. D. Lobel (London: Scolar Press, 1975), 2:7, 9.

17. Quotation in M. Kilian Hufgard, *Saint Bernard of Clairvaux: A Theory of Art Formulated from His Writings and Illustrated in Twelfth-Century Works of Art,* Medieval Studies, no. 2 (Lewiston, N.Y.: E. Mellen Press, 1989), 73. Peter Fergusson, "The First Architecture of the Cistercians in England and the Work of Abbot Adam of Meaux," *JBAA* 136 (1983): 82–86; Christopher Brooke, "St. Bernard, the Patrons, and Monastic Planning," in *Cistercian Art and Architecture in the British Isles,* ed. Christopher Norton and David Parks (Cambridge: Cambridge Univ. Press, 1986), 16–23.

18. The figure is based on my own unofficial list. More sites will undoubtedly be discovered in the coming years, and I may have missed a few that are already known. I am including only sites with probable conduit intake systems; if I were to include sites with elaborate drainage systems and river diversions, the total would be considerably higher. The estimate of 130 houses includes monks, nuns, regular canons, canonesses, and mendicant friars, but not hospitals, churches, or non-monastic cathedrals.

19. The monastic statistics are based on David Knowles and R. Neville Haddock, *Medieval Religious Houses in England and Wales,* 2d ed. (London: Longman, 1971). The numbers of religious are based on the maximum known figure for each house, though with servants and lay brethren, the population of a house could more than double.

20. The 50 percent rate for the Bonhommes is probably a statistical fluke, as the order had only two houses in England.

21. Roberta Gilchrist, *Gender and Material Culture: The Archaeology of Religious Women* (London: Routledge, 1994), 22–25, 42–44, 68, 71, 73; Sally Thompson, *Women Religious: The Founding of English Nunneries after the Norman Conquest* (Oxford: Clarendon Press, 1991), 12–13, 94–95.

22. Kosch, "Wasserbaueinrichtungen," 104.

23. The date of the conduit depends on the dating of the (undated) conduit grants in Godstow's register. Clark dates the conduit charters to c. 1135, on the grounds that the grantor, Robert son of Vincent, Lord of Wytham, was the same Robert of Wytham who was present at the consecration of the church in 1138. Robert's son, however, also seems to have been named Vincent, and this Vincent's son was another Robert (all three are mentioned in a charter of 1160). If the conduit grantor was the younger Robert, a date in the later twelfth century would be likely, which would coincide with Henry II's patronage of the abbey. Andrew Clark, ed., *The English Register of Godstow Nunnery, near Oxford,* pt. 1, EETS, orig. ser., no. 129 (London, 1905), 28–29, 44–46, 49; Thompson, *Women Religious,* 168.

24. *VCH Hampshire,* 2:133. Benoit and Wabont, "Mittelalterliche Wasserversorgung," 212–220; Monique Wabont, *Maubuisson au fil de l'eau: Les réseaux hydrauliques de l'abbaye du XIIIe au XVIIIe siècle,* Notice d'archéologie du Val-d'Oise, no. 3 (Saint-Ouen-l'Aumone: Service Départmental d'Archéologie du Val-d'Oise, 1992); Thompson, *Women Religious,* 12–13, 167–169, 182; Michel Parisse, "Die Frauenstifte und Frauenklöster in Sachsen vom 10. bis zur Mitte des 12. Jahrhunderts," in *Die Salier und das Reich,* ed. Stefan Weinfurter (Sigmaringen, Germany: Jan Thorbecke Verlag, 1992), 2:483; Kenneth H. Rogers, ed., *Lacock Abbey Charters,* Wiltshire Record Society, no. 34 (Devizes, Eng.: Wiltshire Record Society, 1978), 25; Harold Brakspear, "Lacock Abbey, Wilts," *Archaeologia* 57, no. 1 (1900): 125–158; David Sturdy, "The Medieval Nunnery at Clerkenwell," *LA* 2, no. 9 (1974): 217–220; F. A. Gasquet, *English Monastic Life* (London: Methuen, 1904), 173; Gilchrist, *Gender and Material Culture,* 83, 113, 115. My thanks to Kenneth Qualmann and the Winchester Museums Service for permitting me to examine the Nunnaminster excavation records.

25. John F. Benton, ed., *Self and Society in Medieval France: The Memoirs of Abbot Guibert of Nogent* (Toronto: Univ. of Toronto Press, 1984), 60, 62 n. Letter referring to Stephen de Len quoted in H. H. E. Craster, *The Parish of Tynemouth, a History of Northumbria,* vol. 8 (Newcastle-upon-Tyne: Reid, 1907), 76. Jacob Langebek, ed., *Scriptores Rerum Danicarum Medii Aevi,* 9 vols. (Hafniae: F. C. Godiche, 1772–1878) 6:53–54.

26. J. Patrick Greene, *Medieval Monasteries* (Leicester: Leicester Univ. Press, 1992), 129–132.

27. Christopher Thomas, Barney Sloane, and Christopher Phillpotts, *Excavations at the Priory and Hospital of St. Mary Spital, London* (London: Museum of London, 1997), 100; Sturdy, "The Medieval Nunnery at Clerkenwell," 217–220.

28. A. Hamilton Thompson, ed., *Visitations of Religious Houses in the Diocese of Lincoln,* 3 vols., Lincoln Record Society, nos. 7, 14, 21 (Horncastle: Lincoln Record Society, 1913–29) 2:60–63, 114–115, emphasis mine. Peter Heath, ed., *Bishop Geoffrey Blythe's Visitations, 1515–1525,* Collections for a History of Staffordshire, 4th ser., no. 7 ([Oxford]: Staffordshire Record Society, 1973), 2.

29. Vitruvius, *On Architecture,* trans. Frank Granger, 2 vols., Loeb Classical Library, no. 251 (Cambridge: Harvard Univ. Press, 1970), introduction to vol. 1; Sextus Julius Frontinus, *The Stratagems and the Aqueducts of Rome,* ed. Mary B. McElwain, trans. Charles E. Bennett and Clemens Herschel, Loeb Classical Library, no. 174 (Cambridge: Harvard Univ. Press, 1980), xxx–xxxii.

30. Klaus Grewe, *Aquädukt-Marmor: Kalksinter der römischen Eifelwasserleitung als Baustoff des Mittelalters* (Stuttgart: Wittwer, 1992).

31. *Cal.Cl.R. 1231–1234,* 530–531; *LMPME,* 65; E. R. Macpherson and E. G. J. Amos, "The Norman Waterworks in the Keep of Dover Castle," *Arch.Cant.* 43 (1931): 166–172; Giusta Nicco Fasola, *La Fontana di Perugia* (Rome: Libreria dello Stato, 1951), 7–11, 55–63.

32. Philip Burnham, *Cultural Life at Papal Avignon* (Ph.D. diss., Tufts Univ., 1992; Ann Arbor, Mich.: University Microfilms), 384–385, 412–413; Paul Benoit, "Le plomb dans le bâtiment en France à la fin du Moyen Age: L'apport des comptes de construction et de réparation," in *Pierre & métal dans le bâtiment au Moyen Age,* ed. Odette Chapelot and Paul Benoit, Recherches d'histoire et de sciences sociales, no. 11 (Paris: Editions de l'Ecole des Hautes Etudes en Sciences Sociales, 1985), 345; Patrice Beck, "Fontaines et fontainiers des ducs de Bourgogne," *MEFRM* 104, no. 2 (1992): 495–506.

33. Duccio Balestracci, "Li lavoranti non cognosciuti," *Bullettino senese di storia patria* 82–83 (1977): 107–111; Duccio Balestracci, "L'acqua a Siena nel Medioevo," in *Ars et ratio: Dalla torre di Babele al ponte di Rialto,* ed. Jean-Claude Maire Vigueur and Agostino Paravicini Bagliani (Palermo: Sellerio, 1990), 25–27. Audrey M. Erskine, ed. and trans., *The Accounts of the Fabric of Exeter Cathedral, 1279–1353,* 2 vols., Devon and Cornwall Record Society, n.s., nos. 24, 26 (Torquay, Eng.: Devonshire Press, 1981–1983), 2:276–291; *LMPME,* 66, 69.

34. The eleventh-century "Descriptio Farvensis Monasterii" (which is more likely to be a description of Cluny than of Farfa) gives the latrine dimensions of 70 feet long by 23 feet wide, with 45 seats. Wolfgang Braunfels, *Monasteries of Western Europe: The Architecture of the Orders,* trans. Alastair Laing (London: Thames & Hudson, 1972), 238. Christ Church, Canterbury, had a 55-seat "necessarium." Bond, "Urban Monastery," 70; Lanfranc, Archbishop of Canterbury, *The Monastic Constitutions of Lanfranc,* ed. and trans. David Knowles (London: T. Nelson & Sons, 1951).

35. A. G. Little, *Studies in English Franciscan History* (Manchester: Manchester Univ. Press, 1917), 14, 223–224.

36. George M. Foster, *Traditional Societies and Technological Change,* 2d ed. (New York: Harper & Row, 1973), 152–160. Richard G. Feachem, "Water Supplies for Low-Income Communities: Resource Allocation, Planning, and Design for a Crisis Situation," in *WWHHC,* 75–95.

37. Susan Reynolds, *An Introduction to the History of English Medieval Towns*

(Oxford: Clarendon Press, 1982), 62–64; Derek J. Keene, *Cheapside before the Great Fire* (London: Economic and Social Research Council, 1985), 19–20; Clifford T. Smith, *A Historical Geography of Western Europe before 1800* (London: Longman, 1978), 303–304; Josiah Cox Russell, *British Medieval Population* (Albuquerque: Univ. of New Mexico Press, 1948), 140–143, 273–278, 285, 303, 307; H. C. Darby, *A New Historical Geography of England before 1600* (Cambridge: Cambridge Univ. Press, 1976), 134, 184, 243.

38. W. H. Bliss, ed., *Calendar of Entries in the Papal Registers Relating to Great Britain and Ireland: Papal Letters I, A.D. 1198–1304* (London: HMSO, 1893), 46, 53. W. H. Rich Jones, ed., *The Register of S. Osmund,* RS, no. 78 (London: Longman, 1884), 2:6; John M. Steane, *The Archaeology of Medieval England and Wales* (Athens: Univ. of Georgia Press, 1984), 128; K. H. Rogers, "Salisbury," in *The Atlas of Historic Towns,* ed. M. D. Lobel, vol. 1 (London: Scolar Press, 1969).

39. B-P, 2:250, 252, 254. Pericle Perali, "L'acquedotto medievale orvietano: Studio storico e topografico," in *La città costruita: Lavori pubblici e immagine in Orvieto medievale,* by Lucio Riccetti (Florence: Le Lettere, 1992), 235–336; Nicco Fasola, *Fontana di Perugia;* L. C. Alloisi, "Acquedotti e mostre d'acqua dal Medioevo al XIX secolo," in *Il trionfo dell'acqua: Acque e acquedotti a Roma IV sec. a.C.-XX sec.: Mostra, 31 ottobre 1986–15 gennaio 1987,* Museo della civiltà romana (Rome: Paleani, 1986), 201; Duccio Balestracci, *I bottini medievali di Siena* (Siena: Alsaba, 1993).

40. William M. Bowsky, *A Medieval Italian Commune: Siena under the Nine, 1287–1355* (Berkeley: Univ. of California Press, 1981), 219–221; Sandra Tortoli, "Per la storia della produzione laniera a Siena nel Trecento e nei primi anni del Quattrocento," *Bullettino senese di storia patria* 82–83 (1977): 220–238; Con.1262, 3.180; Con.1309, 4.10; B-P, 1:340–344; 2:172–173, 280, 291–293; Duccio Balestracci and Gabriella Piccinni, *Siena nel Trecento: Assetto urbano e strutture edilizie* (Florence: Clusf, 1977), 27, 90–91, 158–163.

41. *Cal.Pat.R. 1374–1377,* 324–325; *Cal.Pat.R. 1399–1401,* 461–462; *Cal.Pat.R. 1446–1452,* 43–44; Edward Gillett and Kenneth A. MacMahon, *A History of Hull* (Oxford: Oxford Univ. Press, 1980), 41–43.

42. All infectious waterborne diseases, which are transmitted by ingesting water containing the pathogens, are fecal-oral diseases. Simply improving water quality without changing other sanitation practices may not eliminate them, but improvements in water quantity and availability can result in more frequent washing, which also reduces the prevalence of these diseases. Richard G. Feachem, "Infectious Disease Related to Water Supply and Excreta Disposal Facilities," *Ambio* 6, no. 1 (1977): 55–58; Feachem, "Water Supplies for Low-Income Communities," 81–91.

43. Derek J. Keene, "Rubbish in Medieval Towns," in *Environmental Archaeol-*

ogy in the Urban Context, ed. R. Hall and H. K. Kenward, CBA Research Report no. 43 (London: CBA, 1982), 26–30; *Cal.L-B A,* 219; *Cal.L-B G,* 300.

44. Morris S. Arnold, ed., *Select Cases of Trespass from the King's Courts 1307–1399,* SS, no. 103 (London: Quaritch, 1987), 2:350; *Cal.Pat.R. 1391–1396,* 353; B-P, 2:254; Marjorie Honeybourne, "The Fleet and Its Neighborhood in Early and Medieval Times," *TLMAS* 19 (1947): 51–52.

45. Quotation in Henry Thomas Riley, ed., *Memorials of London and London Life in the XIIIth, XIVth, and XVth Centuries* (London: Longmans, Green & Co., 1868), 265. The Black Death is conventionally identified as bubonic plague, although this identification has recently been challenged. Whatever the disease, it does not appear to have been spread by a waterborne infective agent. Graham Twigg, *The Black Death: A Biological Reappraisal* (London: Batsford, 1984); Ernest L. Sabine, "City Cleaning in Mediaeval London," *Speculum* 12, no. 1 (1937): 19–43; Joshua Trachtenberg, *The Devil and the Jews: The Medieval Conception of the Jew and Its Relation to Modern Antisemitism* (New Haven: Yale Univ. Press, 1943), 101 ff; Salo Wittmayer Baron, *A Social and Religious History of the Jews,* 2d ed. (New York: Columbia Univ. Press, 1967), 11:160 ff.

46. Riley, *Memorials,* 298–299.

47. B-P, 1: "Pianta dimostrativa della città di Siena," 2:21 n. 1, 101; Balestracci and Piccinni, *Siena nel Trecento,* chap. 1 and map 2; Duccio Balestracci, *I bottini: Acquedotti medievali senesi* (Siena: Comune di Siena, 1984), figs. 3, 6. Con.1309, 3.76, 3.307; Magnusson, *Medieval Water Supplies,* table 1, Group IV fountains.

48. B-P, 2:239–240; Balestracci and Piccinni, *Siena nel Trecento,* 65–68, 145–149, 165–175; S. De Colli, ed., *Libri dell'entrata e dell'uscita del comune di Siena: Detti della Biccherna, Reg. 26 (1257 secondo semestre),* Pubblicazioni degli archivi di Stato, no. 42 (Rome: Ministero dell'interno, 1961), 118–119.

49. Quotation in Judith Hook, *Siena: A City and Its History* (London: H. Hamilton, 1979), 156–157. B-P, 2:284. *Operaio dell'acqua* means roughly "master of the waterworks."

50. Quotation in Riley, *Memorials,* 105–106. E. K. Chambers, *The Medieval Stage* (Oxford: Clarendon Press, 1903), 2:166–173. Lydgate's verses on the ceremonial entry of Henry VI into London in 1432 describe pageants at several conduits. Charles Lethbridge Kingsford, ed., *Chronicles of London* (Oxford: Clarendon Press, 1905), 106–112.

51. Quotation in H. E. Savage, ed., *The Great Register of Lichfield Cathedral Known as Magnum Registrum Album.* Collections for a History of Staffordshire, William Salt Archaeological Society (Kendal: T. Wilson, 1926), 317. According to a writ issued May 17, 1244, Edward, son of Odo, was to receive £1,949 13s. 5½d., which he had spent on a new room, the making of the conduit, and other works at the palace. *Cal.Pat.R. 1232–1247,* 430; J. G. Noppen, "Building by Henry III and

Edward, Son of Odo," *Ant.J.* 29 (1949): 13–25; William Barclay Parsons, *Engineers and Engineering in the Renaissance* (Cambridge: MIT Press, 1967), 241; *VCH Staffordshire,* 14:95.

52. Henry F. Berry, "The Water Supply of Ancient Dublin," *Journal of the Royal Society of Antiquaries of Ireland* 21 (1890–91): 571; *Cal.Dublin,* nos. 59, 63, 71, 80, 93.

53. *Cal.L-B L,* 160.

54. Arnold, *Select Cases of Trespass,* 355–357.

55. Ernest L. Sabine, "Latrines and Cesspools of Mediaeval London," *Speculum* 9, no. 3 (1934): 312.

56. Parsons, *Engineers and Engineering,* 240; Jackson, "Inception of the Dodder Watersupply," 128–141; H. W. Gidden, ed., *The Book of Remembrances of Southampton,* Southampton Record Society, no. 28 (Southampton: Cox & Sharland, 1928), 2:14–16; George Oliver, ed., *Monasticon Dioecesis Exoniensis* (London: Longman, Brown, Green & Longmans, 1846), 115; *Cal.Pat.R. 1340–1343,* 351; *Cal.Pat.R. 1343–1345,* 412, 496.

57. E. W. W. Veale, ed., *The Great Red Book of Bristol: Text,* 2 vols., Bristol Record Society, nos. 4, 8 (Bristol: Bristol Record Society, 1933–38), 2:191–195; H. A. Cronne, ed., *Bristol Charters 1378–1499,* Bristol Record Society, no. 11 (Bristol: Bristol Record Society, 1946), 188–191; Robert Hall Warren, "Some Additional Ecclesiastical Seals of Bristol," *Proceedings of the Clifton Antiquarian Club* 3, no. 3 (1896): 195–203.

58. John Speed, *The History and Antiquity of Southampton,* Southampton Record Society, no. 5 (Southampton: Cox & Sharland, 1909), 25–31; W. H. Stevenson, ed., *Calendar of the Records of the Corporation of Gloucester* (Gloucester: J. Bellows, 1893), doc. 1112.

59. Daniela Monacchi, "L'acquedotto Formina di Narni," *Bollettino d'Arte* 39–40 (1986): 123–142; Alloisi, "Acquedotti e mostre d'acqua," 201; M. W. Frederiksen and J. B. Ward-Perkins, "The Ancient Road Systems of the Central and Northern Ager Faliscus," *PBSR,* n.s., 12 (1957): 104; Giovanni Cecchini, ed., *Il Caleffo Vecchio del comune di Siena,* 4 vols., Fonti di Storia Senese (Florence: L. S. Olschki , 1931–84), vol. 1, nos. 169, 177, 181, 188.

CHAPTER 2. RESOURCE ACQUISITION

1. Rosalind Ransford, ed., *The Early Charters of the Augustinian Canons of Waltham Abbey, Essex 1062–1230* (Woodbridge: Boydell Press, 1989), nos. 400–412.

2. Jim Gould, "The Twelfth-Century Water Supply to Lichfield Close," *Ant.J.* 56 (1976): 75–76; Robert Willis, *The Architectural History of the University of Cambridge and of the Colleges of Cambridge and Eton* (Cambridge: Cambridge Univ. Press, 1886), 2:427, 678–680.

3. Ransford, *Early Charters,* nos. 408–411, 413.

4. C. D. Ross and Mary Devine, eds., *The Cartulary of Cirencester Abbey of Gloucester,* 3 vols. (London: Oxford Univ. Press, 1964–77); for early-thirteenth-century grants, see vol. 1, nos. 235/284–238/287; for later-thirteenth-century grants, vol. 3, nos. 288 and 289. Stow's date of 1285 for London's Great Conduit, though often quoted, is certainly too late. It is referred to as the conduit in Saint Mary Colechurch in West Chepe in 1261, and several references to the conduit in Chepe date to the 1270s. Possibly the London workmen who were attempting to restore the flow of water by cleaning out a blocked conduit pipe in 1256 were engaged in maintenance work on the Great Conduit—if so, it can be presumed to have been completed and in operation by this date. *Cal.L-B A,* 14–15; Thomas Rymer, comp., *Foedera, conventiones, literae et cujuscunque generis acta publica inter reges Angliae,* 3d ed. (Farnborough, Eng.: P. Gregg, 1967), 11:30; William Stubbs, ed., *Chronicles of the Reigns of Edward I and Edward II,* RS, no. 76 (London: Longman, 1882), 1:44; John Stow, *A Survey of London,* ed. Charles L. Kingsford, 2 vols. (Oxford: Clarendon Press, 1908), 1:17; Alfred Stanley Foord, *Springs, Streams, and Spas of London* (London: T. F. Unwin, 1910), 255; Henry Thomas Riley, ed., *Memorials of London and London Life in the XIIIth, XIVth, and XVth Centuries* (London: Longmans, Green & Co., 1868), 14; Matthew Paris, *Matthaei Parisiensis, Monachi Sancti Albani, Chronica Majora,* ed. Henry Richards Luard, RS, no. 57 (London: Longman, 1880), 5:600. On the Southampton conduit, see *Cal.Pat.R. 1281–1292,* 365; *Cal.Pat.R. 1327–1330,* 12.

5. Rymer, *Foedera,* 11:30.

6. Francis B. Bickley, ed., *A Calendar of Deeds (Chiefly Relating to Bristol)* (Edinburgh: T. & A. Constable, 1899), no. 1; H. C. M. Hirst, "Redcliffe Conduit, Bristol, and Robert de Berkeley," *TBGAS* 46 (1924): 359–360; Charles Lethbridge Kingsford, *The Grey Friars of London: Their History with the Register of Their Convent and an Appendix of Documents,* British Society of Franciscan Studies, no. 6 (Aberdeen, Scotland: Aberdeen Univ. Press, 1915), 158–159; J. S. Brewer, ed., *Monumenta Franciscana,* RS, no. 4 (London: Longman, 1858), 1:509–511.

7. Ransford, *Early Charters,* nos. 400–412. The perch was a linear unit of measure, theoretically equivalent to 16½ feet (5½ yards). As with many medieval units of measure, in practice there was considerable local variation.

8. Ross and Devine, *Cartulary of Cirencester Abbey,* 1:227–228, nos. 237/286 and 238/287.

9. Helen M. Cam, *The Hundred and the Hundred Rolls: An Outline of Local Government in Medieval England* (London: Methuen & Co, 1930), 78; *Cal.Pat.R. 1272–1281,* 165; *Cal.Pat.R. 1281–1292,* 75, 442.

10. The jurors' task may have been limited to selecting a source for the city's supply from the stretch of the Poddle south of Saint Thomas's Abbey. *Cal.Dublin,* no. 24; Valentine Jackson, "The Inception of the Dodder Watersupply," in *Medieval*

Dublin: The Making of a Metropolis, ed. Howard Clarke (Dublin: Irish Academic Press, 1990), 128–141.

11. A. R. Martin, *Franciscan Architecture in England,* British Society of Franciscan Studies, no. 18 (Manchester: Manchester Univ. Press, 1937), 219; *VCH Oxford,* 2:111.

12. John Amphlett, ed., *A Survey of Worcestershire by Thomas Habington* (Worcester: Worcestershire Historical Society, 1899), 2:402; R. V. H. Burne, *The Monks of Chester: The History of St. Werburgh's Abbey* (London: SPCK, 1962), 40.

13. Ransford, *Early Charters,* nos. 400–401. Gregorio, di Catino, comp., *Il regesto di Farfa,* ed. Ignazio Giorgi and Ugo Balzani (Rome: Presso la Società, 1883–1914), vol. 2, docs. 99, 100, 101, 107, 114. It is possible, of course, that even the apparent gifts may conceal additional, if unrecorded, temporal inducements or pressures.

14. H. E. Savage, ed., *The Great Register of Lichfield Cathedral Known as Magnum Registrum Album,* Collections for a History of Staffordshire, William Salt Archaeological Society (Kendal: T. Wilson, 1926), no. 528; Martin, *Franciscan Architecture,* 167.

15. *Cal.Pat.R. 1340–1343,* 255; *Cal.Pat.R. 1313–1317,* 128.

16. William Urry, *Canterbury under the Angevin Kings* (London: Athlone Press, 1967), 378. Giovanni Cecchini, ed., *Il Caleffo Vecchio del comune di Siena,* 4 vols., Fonti di Storia Senese (Florence: L. S. Olschki, 1931–84), 2:238–239, 266–269; B-P, 1:149; 2:71; Con.1262, 3.88; Gould, "Lichfield Close," 74; *VCH Stafford,* 14:95; Rymer, *Foedera,* 11:30; A. Morley Davies, "London's First Conduit System: A Topographical Study," *TLMAS,* n.s., 2 (1913): 28–29; *Cal.L-B G,* 210; Exeter Excavation Committee, "Report on the Underground Passages in Exeter," *Proceedings of the Devon Archaeological Exploration Society* 1 (1929–32): 199; *HKW,* 1:551.

17. *Cal.Pat.R. 1281–1292,* 442; Lodovico Zdekauer, *La vita pubblica dei senesi nel Dugento* (Bologna: Forni, 1967), 34–35; Duccio Balestracci and Gabriella Piccinni, *Siena nel Trecento: Assetto urbano e strutture edilizie,* (Florence: Clusf, 1977), 47; Con.1262, 3.182, marginal addition.

18. *Cal.Dublin,* no. 24. De Merton seems to have obtained the land after the license had been issued. George Ormerod, *The History of the County Palatine and City of Chester,* 2d ed. (London: G. Routledge, 1882), 2:176; James Tait, ed., *The Chartulary or Register of the Abbey of St. Werburgh, Chester,* Chetham Society, n.s., no. 79 (Manchester: Chetham Society, 1920), vol. 1, nos. 340–343.

19. B-P, 1:152–154, 159, 161; 2:46, 197–198, 272. A *brachia* was probably an "arm's length"; it was subject to considerable local variation.

20. Quotation in *Cal.Pat.R. 1381–1385,* 537. F. W. Robins, *The Story of Water Supply* (London: Oxford Univ. Press, 1946), 107; *VCH York, E. Riding,* 1:371.

21. Rymer, *Foedera,* 11:30–31; Davies, "London's First Conduit System," 36; *Cal.L-B A,* 14–15.

22. Audrey Woodcock, ed., *Cartulary of the Priory of St. Gregory, Canterbury,* Camden Society, 3d ser., no. 88 (London: Royal Historical Society, 1956), docs. 19, 21, I.5; Robert Willis, "The Architectural History of the Conventual Buildings of the Monastery of Christ Church in Canterbury," *Arch.Cant.* 7 (1868): 182; *Cal.Dublin,* no. 46 ii.

23. Quotation in *Cal.Pat.R. 1281–1292,* 442. Gregorio, di Catino, *Il regesto di Farfa,* vol. 2, docs. 99–101, 107, 114. As other Farfa records indicate, the trees could have included olive trees and fruit trees as well as woodland. L. E. W. O. Fullbrook-Leggatt, "The Water Supplies of the Abbey of St. Peter and the Priory of the Grey Friars, Gloucester, from Robinswood Hill," *TBGAS* 87 (1968): 117–118; *Cal.Glouc.,* no. 962.

24. Stat.1251–52, 1.65; B-P, 2:223.

25. *VCH Stafford,* 14:95; Mary Dormer Harris, ed., *The Coventry Leet Book,* 4 vols., EETS, nos. 134, 135, 138, 146 (London: K. Paul, Trench, Trübner & Co., 1907–13), 1:189–190.

26. *Cal.Pat.R. 1292–1301,* 115–116; *Cal.Pat.R. 1307–1313,* 89; Giusta Nicco Fasola, *La Fontana di Perugia* (Rome: Libreria dello Stato, 1951), 7.

27. Kingsford, *Grey Friars of London,* 48–51, 158–161; Philip Norman, "On an Ancient Conduit-Head in Queen Square, Bloomsbury," *Archaeologia* 56, no. 2 (1899): 258–259; Martin, *Franciscan Architecture,* 201–202.

28. Urry, *Canterbury under the Angevin Kings,* maps 1b, 2b.

29. Rymer, *Foedera,* 11:29–30.

30. Amphlett, *A Survey of Worcestershire by Thomas Habington,* 2:401; *Cal.Dublin,* no. 80.

31. *Cal.Dublin,* no. 46 i; *Cal.Pat.R. 1272–1281,* 165, 299; *Cal.Pat.R 1313–1317,* 398.

32. George Oliver, ed., *Monasticon Dioecesis Exoniensis* (London: Longman, Brown, Green & Longmans, 1846), doc. 20.

33. The conduit of 1325 was the friars' second conduit, apparently replacing one built sometime before 1295. Willis, *Architectural History of the University of Cambridge,* 2:427–430, 678–681; John R. H. Moorman, *The Grey Friars in Cambridge 1225–1538* (Cambridge: Cambridge Univ. Press, 1952), 53–54.

34. The Bishop of Bath and Wells issued them a specific license to enter the chapel yard to mend their pipes. *VCH Somerset,* 2:160; Edmund Hobhouse, ed., *Calendar of the Register of John de Drokensford, Bishop of Bath and Wells* (A.D. *1309–1329*), Somerset Record Society, no. 1 ([London]: Somerset Record Society, 1887), 145; Gould, "Lichfield Close," 76; *Cal.P&M.L. 1364–1381,* 42–43.

35. It seems to have been easier to come up with a theoretical division than it was to put it into practice. The provisions concerning the discharge of water from the reservoir into one pipe for the friars and two for the Abbey have been interlined: the original text provided one pipe for the friars and one larger pipe for the Abbey.

Apparently there was a problem when it came to calculating (or agreeing upon) the relative diameters of the pipes for dividing the water into parts of one-third and two-thirds, and the simpler expedient of three equal pipes was adopted. RCHM, vol. 12, app. 9, 413–414; Fullbrook-Leggatt, "The Water Supplies of the Abbey of St. Peter," 113–115; *Cal.Glouc.*, nos. 962, 966.

36. Riley, *Memorials,* 521; Rymer, *Foedera,* 11:31; Davies, "London's First Conduit System," 26.

37. Jackson, "Inception of the Dodder Watersupply," 133; Davies, "London's First Conduit System," 23–28.

38. *VCH York, North Riding,* 2:538–539; C. James Bond, "Water Management in the Urban Monastery," in *Advances in Monastic Archaeology,* ed. Roberta Gilchrist and Harold Mytum, BAR British Series, no. 227 (Oxford: BAR, 1993), 43–78; William Brown, ed., *Yorkshire Inquisitions,* Yorkshire Archaeological Society Record Series, no. 23 (Leeds: Yorkshire Archaeological Society, 1897), 2:9–11; A. R. Martin, *Franciscan Architecture in England,* British Society of Franciscan Studies, no. 18 (Manchester: Manchester Univ. Press, 1937), 39.

39. Oliver, *Monasticon Dioecesis Exoniensis,* 185.

40. Walter W. Skeat., ed., *Pierce the Ploughman's Crede,* EETS, o.s., no. 30 (London: N. Trübner & Co., 1867), lines 195–216; translation in *The Building of London from the Conquest to the Great Fire,* by John Schofield (London: British Museum, 1984), 72.

CHAPTER 3. DESIGN AND CONSTRUCTION

1. Henry Richards Luard, ed., *Annales Monasterii de Waverleia, Annales Monastici,* vol. 2, RS, no. 36 (London: Longman, Green, Longman, Roberts & Green, 1865), 284–285.

2. References to the waterworks plan of Christ Church, Canterbury, can be found in *LMPME,* 43–58, Waltham Abbey's conduit (plan and accompanying texts) in *LMPME,* 59–70, and the London Charterhouse waterworks plan in *LMPME,* 221–228.

3. "Sancti Bernardi abbatis Clarae-Vallensis vita et res gestae. Libri septem comprehensae," PL, vol. 185, col. 285; English translation in Wolfgang Braunfels, *Monasteries of Western Europe: The Architecture of the Orders,* trans. Alastair Laing (London: Thames & Hudson, 1972), 244. J. Patrick Greene, *Norton Priory: The Archaeology of a Medieval Religious House* (Cambridge: Cambridge Univ. Press, 1989), 34–35.

4. Audrey M. Erskine, ed. and trans., *The Accounts of the Fabric of Exeter Cathedral, 1279–1353,* 2 vols., Devon and Cornwall Record Society, n.s., nos. 24, 26 (Torquay: Devonshire Press, 1981–83).

5. For a well-type conduit-head at Exeter, see Charles Tucker, "Discovery of an Ancient Conduit at St. Sidwell's, near Exeter," *Arch.J.* 15 (1858): 313–317. A conduit house fed by intake adits is known from Saint Augustine's, Canterbury. Paul Bennett, "St. Augustine's Conduit House," *CA* (1987–88): 8–10. The *bottini* (subterranean filtration conduits) at Siena are large adit intakes. Duccio Balestracci, *I bottini medievali di Siena* (Siena: Alsaba, 1993). On Greek and Roman intakes, see A. Trevor Hodge, *Roman Aqueducts and Water Supply* (London: Duckworth, 1992), 27, 69–92.

6. Quotations: Thomas Rymer, comp., *Foedera, conventiones, literae et cujuscunque generis acta publica inter reges Angliae,* 3d ed. (Farnborough: P. Gregg, 1967), 11:30; *Cal.Pat.R. 1330–1334,* 146. *Cal.Pat.R. 1345–1348,* 424; H. E. Salter, ed., *Cartulary of Oseney Abbey,* Oxford Historical Society, no. 97 (Oxford: Clarendon Press, 1934), 4:473–474; *Cal.Chart.R. 1300–1326,* 423–426.

7. *Cal.Pat.R. 1330–1334,* 146.

8. John Brownbill, ed., *The Ledger-Book of Vale Royal Abbey* (Edinburgh: Record Society for the Publication of Original Documents relating to Lancashire and Cheshire, 1914), 226–227; Bennett, "St. Augustine's Conduit House"; Paul Bennett, "St. Augustine's Water Supply," *CA* (1988–89): 13.

9. There are two twelfth-century drawings of the Christ Church water system, both now bound in a psalter of Canterbury Cathedral Priory: the "plan" (Trinity College, Cambridge, MS R.17.1, fols. 284v–285r) and a less elaborate drawing known as the "diagram" (ibid., fol. 286r).

10. British Library, Harley MS 391, fols. 1–6. Mid-thirteenth century. The plan is a diagram, not a scale drawing.

11. In London, the conduit wardens' accounts for 1334 include the cost of making a clay wall around the head of the conduit at Tyburn. *Cal.L-B F,* 28.

12. A thirteenth-century description of the conduit house at Clairvaux stresses its role in protecting the water supply from pollution. "Descriptio positionis seu situationis monasterii Clarae-vallensis," PL, vol. 185, col. 574.

13. W. H. St. John Hope, "Architectural History of Mount Grace Charterhouse," *Yorkshire Archaeological Journal* 18 (1905): 304; Glyn Coppack, *English Heritage Book of Abbeys and Priories* (London: Batsford, 1990), 85–86, fig. 52; Klaus Grewe, "Mount Grace Priory (Yorkshire, GB)," in *WIM,* 264–267; Jim Gould, "The Twelfth-Century Water Supply to Lichfield Close," *Ant.J.* 56 (1976): 73–79; Charles Lethbridge Kingsford, *The Grey Friars of London: Their History with the Register of Their Convent and an Appendix of Documents,* British Society of Franciscan Studies, no. 6 (Aberdeen, Scotland: Aberdeen Univ. Press, 1915), 160–161; Philip Norman, "On an Ancient Conduit-Head in Queen Square, Bloomsbury," *Archaeologia* 56, no. 2 (1899): 251–266; Philip Norman, "Recent Discoveries of Medieval Remains in London," *Archaeologia* 67 (1915–16): 1–26; Philip Norman

and Ernest A. Mann, "On the White Conduit, Chapel Street, Bloomsbury, and Its Connexion with the Grey Friars' Water System," *Archaeologia* 61 (1909): 347–356.

14. W. H. St. John Hope, "The London Charterhouse and Its Old Water Supply," *Archaeologia* 58, no. 1 (1901): 293–312; *Cal.Wells,* 433. Bennett, "St. Augustine's Conduit House"; Bennett, "St. Augustine's Water Supply."

15. W. H. St. John Hope and Harold Brakspear, "The Cistercian Abbey of Beaulieu, in the County of Southampton," *Arch.J.* 63 (1906): 87; Hope, "Mount Grace Charterhouse," 304; Bennett, "St. Augustine's Conduit House," 8–10; *Cal.Wells,* 433.

16. *Cal.Wells,* 433; Hope, "London Charterhouse," 301–303.

17. Salter, *Oseney Abbey,* 473–4; *Cal.Chart.R. 1300–1326,* 424; *Cal.Pat.R. 1330–1334,* 146; Gould, "Lichfield Close," 77; *Cal.Wells,* 433; Norman and Mann, "On the White Conduit," 350–351; Norman, "Recent Discoveries," 21.

18. A calculation of the total capacity of a system would depend on additional factors such as the size of pipes, the rate of flow, and the rate at which the collection cisterns could be refilled.

19. B-P, vol. 1, chap. 3; Balestracci, *I bottini,* 44–45.

20. In Germany stone pipe sections were used as *spolia* in the fourteenth-century castle wall at Miltenberg, but the date of the pipes themselves remains a mystery. An iron pipe, dated to 1455, is known from Schloss Dillenburg. Klaus Grewe, "Wasserversorgung und -entsorgung im Mittelalter: Ein technikgeschichtlicher Überblick," in *WIM,* 37–40.

21. In contrast to closed-pipe flow, open-channel flow is characterized by the existence of a free water surface, where the pressure is atmospheric.

22. E. W. W. Veale, ed., *The Great Red Book of Bristol: Text,* 2 vols., Bristol Record Society, nos. 4, 8 (Bristol: Bristol Record Society, 1933–38), 1:114–119; Erskine, *Fabric of Exeter Cathedral,* 2:274; Henry Thomas Riley, ed., *Memorials of London and London Life in the XIIIth, XIVth, and XVth Centuries* (London: Longmans, Green & Co., 1868), 321–323. A clove was a unit of weight.

23. Erskine, *Fabric of Exeter Cathedral,* 2:274. From time to time the Exeter fabric accounts record large quantities of lead from as far away as Boston, which came by ship to Topsham, Exeter's main port. A permanent plumbery was not unusual for a large institution. Christ Church, Canterbury, had one near the bell tower. A 1399 inventory of the tools and material in the York Minster plumbery included five fothers of lead, two pounds of tin, weighing scales and weights, a brass plane, two soldering irons, a wood axe, a skimmer, a "podyngiren," a tin iron with two hafts and one chisel, a pair of pincers, two new spoons or ladles and one old one, and two fir ladders (2:xiv, xvii, xxiv, 267–268); John Harvey, *Mediaeval Craftsmen* (London: Batsford, 1975), 78.

24. The date of the Waltham furnace is unknown. It is probable that lead pigs or

scrap lead were often merely remelted at the building site, but there are indications that other processing steps might also have been carried out there. An unstratified chunk of galena and barytes (a lead gangue mineral) from medieval contexts at Kirkstall Abbey suggest that at least small-scale smelting operations were carried out near the monastery. P. J. Huggins, "The Excavation of an Eleventh-Century Viking Hall and Fourteenth-Century Rooms at Waltham Abbey, Essex, 1969–71," *MA* 20 (1976): 98; Stephen Moorhouse and Stuart Wrathmell, *Kirkstall Abbey: The 1950–64 Excavations: A Reassessment,* Yorkshire Archaeology, no. 1 (Wakefield: West Yorkshire Archaeology Service, 1987), 1:150; Erskine, *Fabric of Exeter Cathedral,* 1:70, 2:281.

25. L. F. Salzman, *Building in England down to 1540* (Oxford: Clarendon Press, 1952), 264. Medieval lead seems often to have been fairly thoroughly desilverized. In the nineteenth century, soft desilverized lead was considered to be best for pipes, since pipes made from it resisted greater water pressures than pipes made of admixtures with "inferior lead." The possibility that Roman lead was reused in the Middle Ages cannot be excluded, since the compositions of Roman and medieval lead samples are very similar. R. F. Tylecote, *Metallurgy in Archaeology: A Prehistory of Metallurgy in the British Isles* (London: E. Arnold, 1962), table 39; J. Newton Friend and W. E. Thorneycroft, "The Silver Contents of Specimens of Ancient and Mediaeval Lead," *Journal of the Institute of Metals* 41 (1929): table 1, 114–116; John Percy, *The Metallurgy of Lead Including Desilverization and Cupellation* (London: J. Murray, 1870), 506–507, 470 ff. H. M. Colvin, ed., *Building Accounts of King Henry III* (Oxford: Clarendon Press, 1971), 152, 168; Erskine, *Fabric of Exeter Cathedral,* 1:19, 38, 45, 58, 70, 77, 121; L. F. Salzman, *English Industries of the Middle Ages* (London: Constable, 1913), 51–55.

26. This method of casting (along with the use of stone forms) has been suggested for the casting of lead for Roman pipes. The technique was essentially the same as the one used for casting sheets of roofing lead. A description of casting lead sheets for roofs or water pipes is given in the sixteenth-century treatise *Pirotechnia,* by Vannoccio Biringuccio. The thickness of the lead in the walls of medieval pipes ranges from one-eighth to one-half inch (and is consistent with the thickness of lead sheets used for roofing). Cleaning would remove surface impurities and nonmetallic oxides; hammering (whether of the lead sheet or of the rolled pipe on the mandrel) would have the advantage of reducing crystal size. At Exeter an iron hammer was purchased for making lead pipes in 1310–11. Tylecote, *Metallurgy in Archaeology,* 95; Jerome O. Nriagu, *Lead and Lead Poisoning in Antiquity* (New York: Wiley, 1983), 242; Vannoccio Biringuccio, *The Pirotechnia of Vannoccio Biringuccio,* trans. Cyril Stanley Smith and Martha Teach Gnudi (Cambridge: MIT Press, 1959), 376–377; Wladimir W. Krysko, *Lead in History and Art* (Stuttgart: Dr. Riederer-Verlag GmbH, 1979), 61; W. W. Krysko and R. Lehrheuer,

"Metallurgical Investigation of Roman Lead Pipes from Pompeii," *Historical Metallurgy* 10, no. 2 (1976): 53–63; Erskine, *Fabric of Exeter Cathedral,* 1:58.

27. Cast lead generally has large, columnar crystals, which are very sensitive to deformation. Since the recrystallization temperature of lead is below room temperature, any deformation or hammering produces a more finely grained structure. For pipes a fine-grained crystal structure is advantageous, since it is more resistant to corrosion. Whatever the original cross section of lead pipes, over time the internal pressure of the water will deform the lead, tending to make the pipes round. This method of pipe manufacture seems to be generally consistent with Roman techniques. Various other means of making the casting trough for the join have been suggested, such as using the wooden mandrel to form the bottom of the trough or using strips of wood coated in a refractory material such as clay to form the sides. Krysko and Lehrheuer, "Metallurgical Investigation of Roman Lead Pipes," 53–63; R. F. Tylecote, "The Behaviour of Lead as a Corrosion Resistant Medium Undersea and in Soils," *JAS* 10, no. 4 (1983): 397–409; Krysko, *Lead in History,* 61–62; Moorhouse and Wrathmell, *Kirkstall Abbey,* 142; Hodge, *Roman Aqueducts,* 308–315, 469 n. 30.

28. The Romans had three methods of joining lead pipes: using autogenous solder, using lead-tin alloys or soft solders, and mashing the edges of the sheet together and fusing them by drawing a basin of hot coals along the length of the join. Under certain soil conditions, tin undergoes an allotropic change that causes the solder to crumble. Nevertheless, it is possible to use lead-tin solders for water pipes, as the modern practice of wiped soldered joints proves. A medieval lead pipe found at Horsham St. Faith Priory was tested by milliprobe. The resulting quantitative analysis of the metal found "no significant differences" in the composition of the pipe and the seam. The seam in the Exeter pipe was made of molten lead, not a lead-tin solder. A lead pipe from Rievaulx had been welded. Nriagu, *Lead and Lead Poisoning,* 242; Krysko, *Lead in History,* 60–62; Ernesto Paparazzo, "Surface and Interface Analysis of a Roman Lead Pipe 'Fistula': Microchemistry of the Soldering at the Join, as Seen by Scanning Auger Microscopy and X-ray Photoelectron Spectroscopy," *Applied Surface Science* 74 (1994): 61–72; David Sherlock, "Discoveries at Horsham St. Faith Priory, 1970–1973," *Norfolk Archaeology* 36, no. 3 (1976): 219–223; Exeter Excavation Committee, "Report on the Underground Passages in Exeter," *Proceedings of the Devon Archaeological Exploration Society* 1 (1929–32): 200; Friend and Thorneycroft, "Silver Contents," 114; Lynn Thorndike, *History of Magic and Experimental Science* (New York: Columbia Univ. Press, 1923), 2:392; Vincent de Beauvais, *Speculum quadruplex; sive, Speculum majus: Naturale, doctrinale, morale, historiale* (Graz: Akademische Druck-u. Verlagsanstalt, 1964), 7.37.

29. Friend and Thorneycroft, "Silver Contents," 115; *Cal.Pat.R. 1446–1452,* 45; *HKW,* 1:550 n. 4; Erskine, *Fabric of Exeter Cathedral,* 2:280–281; Exeter Excavation

Committee, "Report on the Underground Passages," 200; Salzman, *Building in England,* 277; RCHM, vol. 12, app. 9, 423. Building accounts sometimes include payments not only for tin but also for tallow and/or lard for soldering lead. Tallow is still used as a flux when making wiped soldered joints for lead pipes. Solder and flux formulas are given in the Mappae Clavicula and by Theophilus. Diane Lee Carroll, "Antique Metal-Joining Formulas in the Mappae Clavicula," *Proceedings of the American Philosophical Society* 125, no. 2 (1981): 91–103; Theophilus, *The Various Arts,* ed. and trans. C. R. Dodwell (London: T. Nelson, 1961).

30. Tylecote, *Metallurgy in Archaeology,* 95; Coppack, *Abbeys and Priories,* 91; Exeter Excavation Committee, "Report on the Underground Passages," 200.

31. The eight-inch pipes appear in the narrative after Laurence had made a second set of pipes, presumably to the new specifications. It would probably have been possible to enlarge the diameters of the seven-inch pipes by hammering them (Krysko and Lehrheuer suggest that Roman pipes were hammered while on the mandrel to enlarge their diameters), but Laurence seems to have initially set the smaller pipes aside and then used them later as the conduit approached the abbey. Krysko and Lehrheuer, "Metallurgical Investigation of Roman Lead Pipes," 54.

32. Huggins, "Waltham Abbey," 117; Francis B. Bickley, ed., *A Calendar of Deeds (Chiefly Relating to Bristol)* (Edinburgh: T. & A. Constable, 1899), no. 1; Grewe, "Wasserversorgung und -entsorgung," 33; *Cal.Dublin,* nos. 46 ii., 59, 63, 71, 80, 93; *Cal.Pat.R. 1232–1247,* 430; Robert Hall Warren, "Some Additional Ecclesiastical Seals of Bristol," *Proceedings of the Clifton Antiquarian Club* 3, no. 3 (1896): 195–203.

33. Vitruvius, *On Architecture,* trans. Frank Granger, 2 vols., Loeb Classical Library, no. 251 (Cambridge: Harvard Univ. Press, 1970), 8.6.4; Caius Plinius Secundus, *Natural History,* trans. H. Rackham, 10 vols., Loeb Classical Library (Cambridge: Harvard Univ. Press, 1938–63), 31.57–58; Sextus Julius Frontinus, *The Stratagems and the Aqueducts of Rome,* ed. Mary B. McElwain, trans. Charles E. Bennett and Clemens Herschel, Loeb Classical Library, no. 174 (Cambridge: Harvard Univ. Press, 1980), 1.24–63; H. Fahlbusch, "Über Abflussmessung und Standardisierung bei den Wasserversorgungsanlagen Roms," in *Wasserversorgung im Antiken Rom,* Frontinus-Gesellschaft, Geschichte der Wasserversorgung, no. 1 (Munich: R. Oldenbourg, 1982), 139–144.

34. Veale, *Great Red Book of Bristol,* 1:115. Vitruvius, *On Architecture,* 8.6.4.

35. Similar jointing techniques were used for Roman pipes. E. R. Macpherson and E. G. J. Amos, "The Norman Waterworks in the Keep of Dover Castle," *Arch.Cant.* 43 (1931): 166–172; Hodge, *Roman Aqueducts,* 315; Moorhouse and Wrathmell, *Kirkstall Abbey,* 141–142; Ernest Woolley, "St. Alban's Abbey: Excavations on the Site of the Great Cloister and Adjacent Buildings, 1924," *Transactions of the St. Alban's and Hertfordshire Architectural and Archaeological Society* (1926):

181–189; Coppack, *Abbeys and Priories,* 91; Glyn Coppack, *Abbeys: Yorkshire's Monastic Heritage* (n.p.: English Heritage, 1988), 18, no. 100.

36. Joseph Thomas Fowler, ed., *Extracts from the Account Rolls of the Abbey of Durham,* 3 vols., Surtees Society, nos. 99, 100, 103 (Durham: Andrews & Co., 1898–1901), 2:515, 537; 3:578, 580, 584, 611, 649, 697, 734; Hodge, *Roman Aqueducts,* 470 n. 39; James Raine, ed., *The Durham Household Book; or The Account of the Bursar of the Monastery of Durham from Pentecost 1530 to Pentecost 1534,* Surtees Society, no. 18 (London: J. B. Nichols & Son, 1844), 334.

37. Erskine, *Fabric of Exeter Cathedral,* 276–284; Moorhouse and Wrathmell, *Kirkstall Abbey,* 141–142, no. 320; Coppack, *Abbeys: Yorkshire's Monastic Heritage,* no. 100; Harold Brakspear, *Waverley Abbey* (London: Surrey Archaeological Society, 1905), 42.

38. All three of these types of earthenware pipes were known in the Roman period. The ancient Greeks and Romans also used other varieties of ceramic pipe. Excavations at Fontenay have produced square earthenware pipes with circular bores. Hodge, *Roman Aqueducts,* 111–117; Jean Therasse, "A propos des tuyaux de poterie dans les aqueducs romains," *Les Études Classiques* 46 (1978): 127–132; Paul Benoit and Monique Wabont, "Mittelalterliche Wasserversorgung in Frankreich. Eine Fallstudie: Die Zisterzienser," in *WIM,* 207–211, fig. 21.

39. Plain tapered pipes were thought to be earlier than flanged pipes, and until recently neither type was thought to antedate the thirteenth century (at least in Britain). It was therefore assumed that the introduction of ceramic water pipes occurred after lead pipes were well established. At Kirkham Priory, however, both plain and flanged earthenware pipes were found in a context that was sealed by the 1160–70 phase of the church. If this date is correct, the Kirkham pipes are about the same age as the Christ Church, Canterbury, water system, which is one of the earliest piped systems in Britain. On the Continent, earthenware pipes at Harzburg are thought to date to the period 1065–74. Grace Briscoe and G. C. Dunning, "Medieval Pottery Roof-Fittings and a Water-Pipe Found at Ely," *Proceedings of the Cambridge Antiquarian Society* 6 (1967): 87–89; P. J. Drury, "Chelmsford Dominican Priory: The Excavation of the Reredorter, 1973," *Essex Archaeology and History* 6 (1974): 81 n. 42; Coppack, *Abbeys and Priories,* 94–95, fig. 61; Ralf Busch, "Die Harzburg in Bad Harzburg, Niedersachsen," in *WIM,* 268–271; Glyn Coppack, "Two Late Medieval Pipe-Drains from Thetford Priory," *PSIAH* 33 (1976): 88–90; John P. Allan, *Medieval and Post-Medieval Finds from Exeter, 1971–1980,* Exeter Archaeological Reports, no. 3 (Exeter: Exeter City Council and the Univ. of Exeter, 1984), figs. 49, 77, 130.

40. Roman Malinowski, "Ancient Mortars and Concretes: Aspects of their Durability," *History of Technology* 7 (1982): 97–99; Roman Malinowski, "Einige

Baustoffprobleme der Antiken Äquadukten," in *Journeés d'études sur les aqueducs romains,* ed. J.-P. Boucher (Paris: Société d'Edition "Les Belles Lettres," 1983), 252; Briscoe and Dunning, "Medieval Pottery Roof-Fittings," 86; Paul Bennett, "Rescue Excavations in the Outer Court of St. Augustine's Abbey, 1983–4," *Arch.Cant.* 103 (1986): 101; Drury, "Chelmsford Dominican Priory," 78; Coppack, "Two Late Medieval Pipe-Drains," 88; André Guillerme, "Puits, aqueducs et fontaines: L'alimentation en eau dans les villes du nord de la France, Xe–XIIIe siècles," in *L'eau au moyen âge,* Senefiance, no. 15 (Aix-en-Provence, Marseille: Publications du CUER MA, Univ. de Provence, 1985), 191; Rolf Payne, *Drainage and Sanitation* (London: Construction Press, 1982), 56.

41. The largest medieval lead pipe bores are about 3 inches (about 80 cm). Roman terra-cotta pipes are normally about 16 to 30 inches (40 to 75 cm) long, with internal bores of about 2 to 6 inches (about 5–15 cm). Hodge, *Roman Aqueducts,* 113.

42. Coppack, "Two Late Medieval Pipe-Drains," 88.

43. Briscoe and Dunning, "Medieval Pottery Roof-Fittings," 86; Busch, "Die Harzburg," 268–271; Coppack, "Two Late Medieval Pipe-Drains," 88–89; Coppack, *Abbeys and Priories,* 94, fig. 61; Drury, "Chelmsford Dominican Priory," 78; "Some Minor Excavations Undertaken by the Canterbury Archaeological Trust in 1977–78," *Arch.Cant.* 94 (1978): 183–185; Stanley E. West, "Griff Manor House (Sudeley Castle), Warwickshire," *JBAA,* 3d ser., 31 (1968): 90; Allan, *Medieval and Post-Medieval Finds,* fig. 49, nos. 1529–30.

44. Coppack, "Two Late Medieval Pipe-Drains," 88; Briscoe and Dunning, "Medieval Pottery Roof-Fittings," 86; Susan M. Youngs and John Clark, "Medieval Britain in 1981," *MA* 26 (1982): 209. Late medieval earthenware pipes in Germany were salt glazed and fired at high temperatures. Grewe, "Wasserversorgung und -entsorgung," 34.

45. Most individual potteries produced a very restricted range of products. It is probably safe to conclude that earthenware pipes were made in a small minority of potteries. Grewe, "Wasserversorgung und -entsorgung," 34–35; John Musty, "Medieval Pottery Kilns," in *Medieval Pottery from Excavations: Studies Presented to Gerald Clough Dunning,* ed. Vera I. Evison, H. Hodges, and J. G. Hurst (London: J. Baker, 1974), 41–65.

46. Briscoe and Dunning, "Medieval Pottery Roof-Fittings," 86; "Some Minor Excavations," 181–185; John Musty, D. J. Algar, and P. F. Ewence, "The Medieval Pottery Kilns at Laverstock, near Salisbury, Wiltshire," *Archaeologia* 102 (1969): 139–143; Michael R. McCarthy and Catherine M. Brooks, *Medieval Pottery in Britain A.D. 900–1600* (Leicester: Leicester Univ. Press, 1988), 259, 450.

47. The evidence for local manufacture is based on the similarity of pipe fabrics

to locally made pottery. Local origins have been suggested for earthenware pipes from Canterbury, Reigate, Exeter, and Coldingham Priory. Bennett, "Rescue Excavations," 101; Youngs and Clark, "Medieval Britain in 1981," 209; Allan, *Medieval and Post-Medieval Finds,* 160, 217; Grace A. Elliot and T. D. Thomson, "Coldingham Priory Excavations-II," *History of the Berwickshire Naturalists Club* 38, no. 2 (1969): 101; Grewe, "Wasserversorgung und -entsorgung," 57.

48. Similar assembly marks are known from two of the Thetford Priory pipes. Stewart Cruden, "Glenluce Abbey: Finds Recovered during Excavations," *Transactions, Dumfriesshire and Galloway Natural History and Antiquarian Society* 29 (1950–51): 178, 185, figs. 21, 22; Coppack, "Two Late Medieval Pipe-Drains," 88–89. As far as I know, no kilns with water-pipe wasters have yet been found in association with earthenware pipe systems.

49. West, "Griff Manor House," 90.

50. Improved field techniques in the last few decades have increased the chances of identifying decayed wooden pipes. Nonetheless, if the wood is entirely decayed and no iron jointing collars were used, the traces left by a wooden pipeline will be difficult to distinguish from a host of other linear features. Grewe, "Wasserversorgung und -entsorgung," 34–36; Niklaus Flüeler, ed., *Stadtluft, Hirsebrei und Bettelmönch: Die Stadt um 1300* (Stuttgart: Theiss, 1992), 274, 373–374; Niklaus Schnitter, *Die Geschichte des Wasserbaus in der Schweiz* (Oberbözberg, Switz.: Olynthus, 1992), 56–57.

51. In Germany pipes were made of oak, spruce, and pine. Grewe, "Wasserversorgung und -entsorgung," 36–38, 57–58; Flüeler, *Stadtluft, Hirsebrei und Bettelmönch,* 363; Hodge, *Roman Aqueducts,* 112; Coppack, *Abbeys and Priories,* 94–95; David R. M. Gainster, Sue Margeson, and Maurice Hurley, "Medieval Britain and Ireland in 1989," *MA* 34 (1990): 228; Salzman, *Building in England,* 441–442. Hope and Brakspear, "Cistercian Abbey of Beaulieu," 144; C. James Bond, "Water Management in the Rural Monastery," in *The Archaeology of Rural Monasteries,* ed. R. Gilchrist and H. Mytum, BAR British Series, no. 203 (Oxford: BAR, 1989), 87; Peter R. V. Marsden, "Archaeological Finds in the City of London, 1963–4," *TLMAS* 21 (1967): 216; Martin Biddle, Lawrence Barfield, and Alan Millard, "The Excavation of the Manor of the More, Rickmansworth, Hertfordshire," *Arch.J.* 116 (1959): 154.

52. Tapered wooden pipes from Braunschweig had internal diameters from 14.2 to 8.2 cm. The Thornholme bore of 11 cm was large by Roman standards. Roman wooden pipes generally had a longitudinal bore of 5–10 cm in diameter. The sixteenth-century wooden pipes at Danzig were thirty-six feet long. Coppack, *Abbeys and Priories,* 95; Grewe, "Wasserversorgung und -entsorgung," 36, 58. Hodge, *Roman Aqueducts,* 112; Marsden, "Archaeological Finds in the City of London," 216.

53. Salzman, *Building in England,* 441–442; Michael Aston, ed., *Medieval Fish,*

Fisheries, and Fishponds in England, BAR British Series, no. 182 (Oxford: BAR, 1988), 1:27.

54. National Tank and Pipe Company, *A Handbook of Wood Pipe Practice* (Portland, Oreg.: National Tank & Pipe Co., 1938), 7–8; Schnitter, *Die Geschichte des Wasserbaus,* 56; Grewe, "Wasserversorgung und -entsorgung," 36.

55. Veale, *Great Red Book of Bristol,* 1:116, Payne, *Drainage and Sanitation,* 124–127; J. B. White, *Wastewater Engineering* (London: E. Arnold, 1978), 6, 30–32.

56. W. H. St. John Hope and J. T. Fowler, "Recent Discoveries in the Cloister of Durham Abbey," *Archaeologia* 58, no. 2 (1903): 446; Fowler, *Account Rolls of the Abbey of Durham,* 2:536; 3:654; Cruden, "Glenluce Abbey," 185; M. O. H. Carver, "Excavations South of Lichfield Cathedral, 1976–77," *TSSAHS* 22 (1982): fig. 5; Bennett, "Rescue Excavations," 101.

57. Matthew Paris, *Matthaei Parisiensis, Monachi Sancti Albani, Chronica Majora,* ed. Henry Richards Luard, RS, no. 57 (London: Longman, 1880), 5:600; Kingsford, *Grey Friars of London,* 158–160; James Craigie Robertson, ed., *Materials for the History of Thomas Becket, Archbishop of Canterbury,* 7 vols., RS, no. 67 (London: Longman, 1875–85), 1:253–254; 2:261–263; John Hayes, "Prior Wibert's Waterworks," *Canterbury Cathedral Chronicle* 71 (1977): 23. It is possible that the depth of the trench was exaggerated in the recounting of the miracle.

58. The gradients of medieval water systems have not been studied in detail (which would require the excavation of long stretches of pipe trenches or channels). When gradients are given in the secondary literature, they are usually mathematical abstractions based on the overall distance and difference in elevation between the source and the terminus. Such theoretical calculations can determine the maximum average gradient for a uniform slope in a direct line, but they shed little light on a system's actual profile. Recent work on Roman aqueduct gradients provides some useful models for future medieval research. Hodge, *Roman Aqueducts,* 178–191; C. G. Henderson, "King William Street Excavation and the Underground Passages," in *Archaeology in Exeter 1983/4* (Exeter: Exeter Museum, 1984), 21–25.

59. Balestracci, *I bottini,* 47; James France, *The Cistercians in Scandinavia,* Cistercian Studies, no. 131 (Kalamazoo, Mich.: Cistercian Publications, 1992), 14–15; Hodge, *Roman Aqueducts,* chap. 7; F. Bluhme, K. Lachmann, and A. Rudorff, eds., *Die Schriften der römischen Feldmesser,* 2 vols. (Hildesheim, Germany: G. Olms, 1967); Thomas F. Glick, "Levels and Levelers: Surveying Irrigation Canals in Medieval Valencia," *Technology and Culture* 9 (1968): 165–180; Frank Prager and Gustina Scaglia, *Mariano Taccola and His Book De Ingeneis,* (Cambridge: MIT Press, 1972), 130–131.

60. Veale, *Great Red Book of Bristol,* 1:115–116.

61. Carver, "Excavations South of Lichfield Cathedral," fig. 5; Moorhouse and

Wrathmell, *Kirkstall Abbey,* 15; S. M. Hirst, D. A. Walsh, and S. M. Wright, *Bordesley Abbey II: Second Report on Excavations at Bordesley Abbey, Redditch, Hereford-Worcestershire,* BAR British Series, no. 111 (Oxford: BAR, 1983), 40–43.

62. John P. Allan, *Exeter's Underground Passages* (Exeter: Exeter City Museums and Art Gallery, 1994); Erskine, *Fabric of Exeter Cathedral,* 2:276–282.

63. Philip Rahtz and Susan Hirst, *Bordesley Abbey, Redditch, First Report on the Excavations 1969–73,* BAR British Series, no. 23 (Oxford: BAR, 1976), 205; Hirst, Walsh, and Wright, *Bordesley Abbey II,* 112; Allan, *Medieval and Post-Medieval Finds,* 91; Bennett, "Rescue Excavations," 101; Moorhouse and Wrathmell, *Kirkstall Abbey,* 9–11.

64. Hope, "London Charterhouse," 306. Bryan Ward-Perkins, *From Classical Antiquity to the Middle Ages: Urban Public Building in Northern and Central Italy,* A.D. 300–850 (Oxford: Oxford Univ. Press, 1984), 136, 189; John Amphlett, ed., *A Survey of Worcestershire by Thomas Habington* (Worcester: Worcestershire Historical Society, 1899), 2:403; C. James Bond, "Water Management in the Urban Monastery," in *Advances in Monastic Archaeology,* ed. Roberta Gilchrist and Harold Mytum, BAR British Series, no. 227 (Oxford: BAR, 1993), 54–55.

65. Bond, "Urban Monastery," 50; Henderson, "King William Street," 24; Kingsford, *Grey Friars of London,* 160; S. W. Ward, *Excavations at Chester: The Lesser Medieval Religious Houses. Sites Investigated 1964–1983,* Grosvenor Museum Archaeological Excavation and Survey Reports, no. 6 (Chester: Chester City Council, 1990), 60; B-P, 1:220–223; Anne Coffin Hanson, *Jacopo della Quercia's Fonte Gaia* (Oxford: Clarendon Press, 1965), 6–7; Pericle Perali, "L'acquedotto medievale orvietano. Studio storico e topografico," in *La città costruita: Lavori pubblici e immagine in Orvieto medievale,* by Lucio Riccetti (Florence: Le Lettere, 1992), 235–336. The use of large-scale inverted siphons in Roman water systems has only recently been widely recognized and studied. A. Trevor Hodge, "Siphons in Roman Aqueducts," *PBSR,* n.s., 51 (1983): 174–221.

66. Hope, "London Charterhouse," 301–311; Salzman, *Building in England,* 270–272; Norman and Mann, "On the White Conduit," 354–356, pl. 46; Kingsford, *Grey Friars of London,* 48–52; *Cal.L-B H,* 326; Riley, *Memorials,* 503–504; R. E. Latham, *Revised Medieval Latin Word-List from British and Irish Sources* (Oxford: Oxford Univ. Press, 1965), 40, 352. In the absence of specific evidence to the contrary, most medieval references to suspirals are probably best interpreted as references to some sort of small tank.

67. Kingsford, *Grey Friars of London,* 160; Norman and Mann, "On the White Conduit," 354; Warwick Rodwell, *Wells Cathedral: Excavations and Discoveries* (Wells: St. Andrew's Press, 1980), 15. In the absence of stopcocks, the mouths of the pipes could simply be plugged. Some tanks seem to have been provided with overflow systems.

68. Grewe, "Wasserversorgung und -entsorgung," 34–35; Martin Biddle, "Excavations at Winchester, 1971. Tenth and Final Interim Report: Part II," *Ant.J.* 55 (1975): 330–332.

69. Rodwell, *Wells Cathedral,* 15, fig. 12; Salzman, *Building in England,* 275.

70. Hope, "London Charterhouse," 301 n.

71. Moorhouse and Wrathmell, *Kirkstall Abbey,* 9–11, 14, figs. 7, 29, 30; Biddle, "Excavations at Winchester, 1971," 332, pl. 41. Cruden, "Glenluce Abbey," 185, fig. 21. The Romans often preferred to use junction boxes rather than strengthen the pipes at bends and junctions. Hodge, *Roman Aqueducts,* 120, 317–320.

72. Biddle, "Excavations at Winchester, 1971," 330–332; Mary Dormer Harris, ed., *The Coventry Leet Book,* 4 vols., EETS, nos. 134, 135, 138, 146 (London: K. Paul, Trench, Trübner & Co., 1907–13), 1:21, 105, 157; 2:549; 3:584.

73. For the various craftsmen employed building conduits and gutters at Westminster Palace, see Colvin, *Building Accounts of King Henry III,* 203–287.

74. Henry Thomas Riley, ed., *Gesta Abbatum Monasterii Sancti Albani a Thoma Walsingham,* RS, no. 28.4 (London: Longmans, Green & Co., 1869), 3:443; Harold Brakspear, "On the Dorter Range at Worcester Priory," *Archaeologia* 67 (1916): 202; Meredith Parsons Lillich, "Cleanliness with Godliness: A Discussion of Medieval Monastic Plumbing," in *Mélanges à la mémoire du père Anselme Dimier,* ed. Benoît Chauvin, vol. 3, no. 5 (Pupillin, Arbois: B. Chauvin, 1982), 144; Hope and Brakspear, "Cistercian Abbey of Beaulieu," 141; Greene, *Norton Priory,* 34–35; John R. Kenyon, *Medieval Fortifications* (Leicester: Leicester Univ. Press, 1990), 131.

75. The cross-sectional area used in the hydraulic equations refers to the part of the channel actually filled with water, not the dimensions of the total channel. Sometimes channels preserve telltale "tidemarks" or incrustations on the walls, which allow an approximate estimate of the usual depth of flow. See, for example, Drury, "Chelmsford Dominican Priory," 50. The basic principles and formulas are lucidly explained in Hodge, *Roman Aqueducts,* 215–226, 349–355.

76. The overall average Roman gradients generally range from 0.15 to 0.3 percent. A slope that rises 1 meter over a distance of 100 meters has a gradient of 1 percent; a rise of 2.5 meters over a distance of 100 meters would be a gradient of 2.5 percent. Hodge, *Roman Aqueducts,* 217–219, 126–147, 205–213; Benoit and Wabont, "Mittelalterliche Wasserversorgung," 189; Ward, *Excavations at Chester,* 44; Brakspear, "On the Dorter Range at Worcester Priory," 202; Harold Brakspear, "Bardney Abbey," *Arch.J.* 79 (1922): 40; Greene, *Norton Priory,* 122–123. Balestracci, *I bottini,* 100–101; Klaus Grewe, "Der Fulbert-Stollen am Laacher See," in *WIM,* 277–281; Heinz Dopsch, "Der Salzburger Almkanal," in *WIM,* 282–286; Armando Schiavo, *Acquedotti romani e medioevali* (Naples: F. Giannini, 1935).

77. Ward, *Excavations at Chester,* 44; Drury, "Chelmsford Dominican Priory," 46, 49.

78. Greene, *Norton Priory,* 122–123, fig. 71; J. Patrick Greene, *Medieval Monasteries* (Leicester: Leicester Univ. Press, 1992), 131; Brakspear, "On the Dorter Range at Worcester Priory," 202.

79. Greene, *Norton Priory,* 75; Jean E. Mellor and T. Pearce, *The Austin Friars, Leicester,* CBA Research Report no. 35 (London: CBA, 1981), 11–17, 24–25, 35.

80. Kenyon, *Medieval Fortifications,* 130–132.

81. G. A. B. Young, "Excavations at Southgate, Hartlepool, Cleveland, 1981–82," *Durham Archaeological Journal* 3 (1987): 22; George Lambrick, "Further Excavations on the Second Site of the Dominican Priory, Oxford," *Oxoniensia* 50 (1985): 151–153; Cyril Stanley Smith and John G. Hawthorne, "Mappae Clavicula: A Little Key to the World of Medieval Techniques," *Transactions of the American Philosophical Society,* n.s., 64, no. 4 (1974): 42, 254; Colvin, *Building Accounts of King Henry III,* 302–304; Erskine, *Fabric of Exeter Cathedral,* 1:25, 53, 55, 56, 65; Nevio Basezzi and Bruno Signorelli, *Gli antichi acquedotti di Bergamo* (Bergamo: Comune di Bergamo, 1992), 107–122; Salzman, *Building in England,* 153–154; Benoit and Wabont, "Mittelalterliche Wasserversorgung," 191; Yosef Porath, "Lime Plaster in Aqueducts: A New Chronological Indicator," *Leichtweiss-Institut für Wasserbau der Technischen Universität Braunschweig: Mitteilungen* 82 (1984); Susan M. Youngs, John Clark, and T. B. Barry, "Medieval Britain and Ireland in 1985," *MA* 30 (1986): 126; Young, "Excavations at Southgate, Hartlepool," 22; Mellor and Pearce, *Austin Friars, Leicester,* 35.; Glyn Coppack, "The Excavation of an Outer Court Building, Perhaps the Woolhouse, at Fountains Abbey, North Yorkshire," *MA* 30 (1986): 55.

82. Quotations in *Cal.Pat.R. 1292–1301,* 288; and *Calendar of the Liberate Rolls Preserved in the Public Record Office,* vol. 4, *1251–1260* (London: HMSO, 1959), 507. *Cal.Cl.R. 1256–1259,* 377–378, 380; Colvin, *Building Accounts of King Henry III,* 196, 288–331.

83. P. J. Tester, "Excavations on the Site of Higham Priory," *Arch.Cant.* 82 (1967): 149; R. Gilyard-Beer, "Fountains Abbey: The Early Buildings, 1132–50," *Arch.J.* 125 (1968): 317; Stuart Wrathmell, *Kirkstall Abbey: The Guest House,* 2d ed. (Wakefield: West Yorkshire Archaeology Service, 1987), 15; Eric J. Boore, "The Lesser Cloister and a Medieval Drain at St. Augustine's Abbey, Bristol," *TBGAS* 107 (1989): 220–221; Colin Platt and Richard Coleman-Smith, *Excavations in Medieval Southampton 1953–1969,* 2 vols. (Leicester: Leicester Univ. Press, 1975), vol. 1, fig. 94; Kenyon, *Medieval Fortifications,* fig. 6.2. On Roman channel junctions and transitions, see Hodge, *Roman Aqueducts,* 118–123.

84. German scholars have recorded sinter in some medieval pipes and channels. The reredorter drain at Chelmsford Dominican Priory had a "thick brown accretion" on the walls, which had built up over a period of years and seems to have been associated with excrement. Grewe, "Wasserversorgung und -entsorgung," 34;

Clemens Kosch, "Wasserbaueinrichtungen in hochmittelalterlichen Konventanlagen Mitteleuropas," in *WIM,* 114; Drury, "Chelmsford Dominican Priory," 50, 80 n. 18; Balestracci, *I bottini,* photographs; B-P, 1:43; 2:362. On sinter in Roman pipes and aqueducts, see H. Fahlbusch, "Maintenance Problems in Ancient Aqueducts," in *Future Currents in Aqueduct Studies,* ed. A. Trevor Hodge, Collected Classical Papers, no. 2 (Leeds: F. Cairns, 1991), 7–14; Klaus Grewe, *Aquädukt-Marmor: Kalksinter der römischen Eifelwasserleitung als Baustoff des Mittelalters* (Stuttgart: Wittwer, 1992).

85. By "primary sediments" I mean those sediments deposited by the flow of water during the working life of a channel. When excavated, many channels also have backfills of secondary material, such as building debris, deposited after the system had gone out of use. David Beard, "The Infirmary of Bermondsey Priory," *LA* 5, no. 7 (1986): 190.

86. Erskine, *Fabric of Exeter Cathedral,* 2:261; P. J. Tester, "Excavations at Boxley Abbey," *Arch.Cant.* 88 (1973): 137; Brakspear, "Bardney Abbey," 40, pl. 8, no. 1.; Young, "Excavations at Southgate, Hartlepool," 22; Hodge, *Roman Aqueducts,* 102–103.

87. Youngs, Clark, and Barry, "Medieval Britain and Ireland in 1985," 184; Greene, *Norton Priory,* 34–36, fig. 20; Kosch, "Wasserbaueinrichtungen," 93; Carolyn Heighway, "Archaeology in the Precinct of Gloucester Cathedral, 1983–5," *Glevensis* 22 (1988): 36; Bond, "Urban Monastery," 49; Tester, "Excavations at Boxley Abbey," fig. 2; Aston, *Medieval Fish,* 27.

88. Henry Summerson, *Medieval Carlisle: The City and the Borders from the Late Eleventh to the Mid–Sixteenth Century,* Cumberland and Westmorland Antiquarian and Archaeological Society, extra ser., no. 25 (1993), 1:163–164. Rob Poulton and Humphrey Woods, *Excavations on the Site of the Dominican Friary at Guildford in 1974 and 1978,* Research Volume of the Surrey Archaeological Society, no. 9 (Guildford, Eng.: Surrey Archaeological Society, 1984), 37, pl. 10; Leslie E. Webster and John Cherry, "Medieval Britain in 1978," *MA* 23 (1979): 248; Tester, "Excavations at Boxley Abbey," 158; Aston, *Medieval Fish,* 99; Greene, *Norton Priory,* 34–35; Peter Somerville-Large, *Dublin: The First Thousand Years* (Belfast: Appletree Press, 1988), 73–74; Urry, *Canterbury under the Angevin Kings,* 186.

89. Vitruvius, *On Architecture,* 8.6.1–2; Hodge, *Roman Aqueducts,* 280 ff.; Jen Dybakjaer Larsen, "The Water Towers in Pompeii," *Anacleta Romana Instituti Danici* 11 (1982): 41–67.

90. Hope, "London Charterhouse," 307–309; Bond, "Urban Monastery," 65; Bond, "Rural Monastery," 88.

91. W. Douglas Caröe, "The Water Tower," *Friends of Canterbury Cathedral Report* 2 (1929): 25–37; Klaus Grewe, "Der Wasserversorgungsplan des Klosters Christchurch in Canterbury (12. Jahrhundert)," in *WIM,* 229–233; Robert Willis,

"The Architectural History of the Conventual Buildings of the Monastery of Christ Church in Canterbury," *Arch.Cant.* 7 (1868): 63–64, 162–163, fig. 7.

92. Moorhouse and Wrathmell, *Kirkstall Abbey,* 9–11, 15; Webster and Cherry, "Medieval Britain in 1978," 250; Biddle, "Excavations at Winchester, 1971," 330–332; Rodwell, *Wells Cathedral,* 14–15, figs. 3–5; Edmund Buckle, "On the Lady Chapel by the Cloister of Wells Cathedral and the Adjacent Buildings," *PSANHS* 40, no. 2 (1894): 35–36, 45–47.

93. Walter H. Godfrey, "English Cloister Lavatories as Independent Structures," *Arch.J.* 106 suppl. (1952): 91–97; Heinrich Grüger, "Cistercian Fountain Houses in Central Europe," in *Studies in Cistercian Art and Architecture,* ed. Meredith P. Lillich, Cistercian Studies Series, no. 69 (Kalamazoo, Mich.: Cistercian Publications, 1984), 2:201–222; Bond, "Urban Monastery," 68. The *Rites of Durham* gives a vivid description of a cloister fountain and a fountain house. Joseph Thomas Fowler, ed., *Rites of Durham,* Surtees Society, no. 107 (Durham: Andrews & Co., 1903), 82–83.

94. Surviving English town conduits are mostly postmedieval, though their lead cisterns and cocks are probably similar to the medieval arrangement. Bond, "Mittelalterliche Wasserversorgung," 171–180; F. W. Robins, *The Story of Water Supply* (London: Oxford Univ. Press, 1946), 129–140.

95. Grewe, "Wasserversorgung und -entsorgung," 56–59.

96. Walter W. Skeat, ed., *Pierce the Ploughman's Crede,* EETS, o.s., no. 30 (London: N. Trübner & Co., 1867), lines 195–196. Grewe, "Wasserversorgung und -entsorgung," 55–56; Thomas Duffus Hardy, ed., *Rotuli Litterarum Clausarum* (London: Record Commission, 1833), 1:140b; *HKW,* 1:549; 3.1:318; "Inventories of the Religious Houses of Shropshire at Their Dissolution," *Transactions of the Shropshire Archaeological and Natural History Society,* 3d ser., 5 (1905): 380; John Stow, *A Survey of London,* ed. Charles L. Kingsford, 2 vols. (Oxford: Clarendon Press, 1908), 1:319.

97. Valerie Horsman and Brian Davison, "The New Palace Yard and Its Fountains: Excavations in the Palace of Westminster 1972–4," *Ant.J.* 69, no. 2 (1989): 291–295; Francesco Mattioli, "Il peperino di Viterbo—una nota storica," in *Peperino & Basaltina* (Viterbo: Consorzio Viterbo-Export, 1986), 33–39; B-P, 1:51–52.

98. Percy H. Reaney, *The Origin of English Surnames* (London: Routledge & Kegan Paul, 1967), 189. James Raine, ed., *Historiae Dunelmensis Scriptores Tres,* Surtees Society, no. 9 (London: J. B. Nichols & Son, 1839), ccccxliii–ccccxlv; Hope and Fowler, "Recent Discoveries," 448–451, 458–460; Giusta Nicco Fasola, *La Fontana di Perugia* (Rome: Libreria dello Stato, 1951), 55–63.

99. On Roman taps and stopcocks, see Hodge, *Roman Aqueducts,* 322–326.

100. G. C. Dunning, "Medieval Bronze Tap-Handles from Lewes and Kirkstall Abbey," *Ant.J.* 48 (1968): 310–311; Moorhouse and Wrathmell, *Kirkstall Abbey,* 22, 121, 132–133; J. T. Micklethwaite, "On a Filtering Cistern of the Fourteenth Cen-

tury at Westminster Abbey," *Archaeologia* 53, no. 1 (1892): 167–168; Martin J. Hicks, "St. Gregory's Priory," *CA* (1988–89): 19; Alison Hicks and Alan Ward, personal communication, 1990; Coppack, *Abbeys and Priories,* 93, fig. 63.

101. Fowler, *Rites of Durham,* 82. Grewe, "Der Wasserversorgungsplan des Klosters Christchurch in Canterbury," 234; Hope, "London Charterhouse," 310. The Charterhouse plan's notes identifies them as "the launderi cok" and "the botery cok."

102. Quotation in Reaney, *The Origin of English Surnames,* 189. Tylecote, *Metallurgy in Archaeology,* 57; *HKW,* 1:549–551; Salzman, *Building in England,* 275–276.

103. *Cal.L-B F,* 29; Salzman, *Building in England,* 275; *HKW,* 1:549.

104. A good model for future studies is the comprehensive analysis of the waterworks at Maubuisson. Monique Wabont, *Maubuisson au fil de l'eau. Les réseaux hydrauliques de l'abbaye du XIIIe au XVIIIe siècle,* Notice d'archéologie du Val-d'Oise, no. 3 (Saint-Ouen-l'Aumone: Service Départmental d'Archéologie du Val-d'Oise, 1992).

105. For example, Fountains Abbey had at least two separate lead-pipe systems, fed by different sources and supplying different parts of the monastic complex. Towns with multiple civic conduits include Bristol, London, Siena, and Viterbo.

CHAPTER 4. ADMINISTRATION AND FINANCE

1. Eleanor Searle, ed. and trans., *The Chronicle of Battle Abbey* (Oxford: Clarendon Press, 1980), 44–45. Eleanor Searle and Barbara Ross, eds., *The Cellarers' Rolls of Battle Abbey: 1275–1513,* Sussex Record Society, no. 65 (Lewes: Sussex Record Society, 1967); Joseph Thomas Fowler, ed., *Extracts from the Account Rolls of the Abbey of Durham,* 3 vols., Surtees Society, nos. 99, 100, 103 (Durham: Andrews & Co., 1898–1901), 3:654; G. W. Kitchin, ed., *Compotus Rolls of the Obedientiaries of St. Swithun's Priory, Winchester, from the Winchester Cathedral Archives,* Hampshire Record Society, no. 5 (London, 1892).

2. Joan Evans, *Monastic Life at Cluny* (London: Oxford Univ. Press, 1931), 71; *VCH York,* 3:275; *VCH Lincoln,* 3:214; Audrey M. Erskine, ed. and trans., *The Accounts of the Fabric of Exeter Cathedral, 1279–1353,* 2 vols., Devon and Cornwall Record Society, n.s., nos. 24, 26 (Torquay: Devonshire Press, 1981–83), 2:282–283, 290.

3. L. F. Salzman, *Building in England down to 1540* (Oxford: Clarendon Press, 1952), 396–397. Colin Platt, *Medieval Southampton: The Port and Trading Community,* A.D. *1000–1600* (London: Routledge & Kegan Paul, 1973), 144; Edmund Hobhouse, ed., *Calendar of the Register of John de Drokensford, Bishop of Bath and Wells (*A.D. *1309–1329),* Somerset Record Society, no. 1 ([London]: Somerset Record Society, 1887), 145.

4. RCHM, vol. 12, app. 9, 424. E. W. W. Veale, ed., *The Great Red Book of*

Bristol: Text, 2 vols., Bristol Record Society, nos. 4, 8 (Bristol: Bristol Record Society, 1933–38) 1:114–119; *VCH Glouc.,* 4:34–35, 62, 262.

5. Giusta Nicco Fasola, *La Fontana di Perugia* (Rome: Libreria dello Stato, 1951), 8, 58–59.

6. *Cal.L-B A,* 162; *Cal.L-B C,* 9; *Cal.L-B D,* 236–237; *Cal.L-B E,* 204–205; Henry Thomas Riley, ed., *Memorials of London and London Life in the XIIIth, XIVth, and XVth Centuries* (London: Longmans, Green & Co., 1868), 77–78, 107, 201–202. On the London Bridge wardens, see Douglas Knoop and G. P. Jones, *The Mediaeval Mason: An Economic History of English Stone Building in the Later Middle Ages and Early Modern Times,* 3d ed. (Manchester: Manchester Univ. Press, 1967), 32–34.

7. *Cal.L-B E,* 204; Mary Dormer Harris, ed., *The Coventry Leet Book,* 4 vols., EETS, nos. 134, 135, 138, 146 (London: K. Paul, Trench, Trübner & Co., 1907–13), 3:584.

8. A "cotiler" or "cotiller" is a cutler; a "coffrer" is a cofferer, that is, a treasurer or a chest maker; a "chaundiler" is a chandler, a candlemaker; an "isemonger" may be an icemonger. *Cal.L-B C,* 9; *Cal.L-B E,* 204–205; *Cal.L-B G,* 2, 11; *Cal.L-B H,* 127–128; Riley, *Memorials,* 264–265.

9. *Cal.L-B G,* 223–224.

10. Stat.1251–52, 1.47–61, 3.145.

11. B-P, 2:72–74.

12. B-P, 2:76, 82, 97, 105–107, 109–111, 113, 115, 124, 199–223.

13. Daniel Waley, *Siena and the Sienese in the Thirteenth Century* (Cambridge: Cambridge Univ. Press, 1991), 59. Con.1262, 3.175; Con.1309, 3.104.

14. B-P, 2:76, 82, 105, 106, 107, 110, 113.

15. There seems to have been a more frequent schedule for cleaning Fonte Branda's *guazatoio:* the statutes call for it to be cleaned twice in winter and four times in summer, but the Biccherna records do not suggest that it was really cleaned this often. B-P, 2:10, 32, 82, 122, 140, 176–177; Con.1262, 3.171, 196; Con.1309, 3.96, 105.

16. B-P, vol. 1, chap. 6; 2:39–40, 137–138, 166–167, 179, 184, 189; William M. Bowsky, *A Medieval Italian Commune: Siena under the Nine, 1287–1355* (Berkeley: Univ. of California Press, 1981), 42–45. Siena was divided into three administrative zones (*terzi*) rather than into four quarters as many other cities were.

17. B-P, 2:38, 193–194, 196, 199–209, 212–217, 221–224.

18. Endres Tucher, *Endres Tuchers Baumeisterbuch der Stadt Nürnberg (1464–1475),* ed. Matthias Lexer, Bibliothek des Litterarischen Vereins in Stuttgart, no. 64 (Amsterdam: Rodopi, 1968).

19. See William M. Bowsky, *The Finance of the Commune of Siena 1287–1355* (Oxford: Clarendon Press, 1970), app. 1, for a table of Biccherna budgets from 1286

to 1355. These show incomes ranging from 26,000 to over 300,000 lire per semester. Overall, the budgets tended to increase with time, but there were striking fluctuations from semester to semester. Bowsky, *Finance of the Commune of Siena,* 222, 244–245; Bowsky, *Medieval Italian Commune,* 10, 191–192; B-P, 1:112–114. The *contado* of Siena refers to the lands (the rural hinterland and its settlements) under the jurisdiction of the commune—Siena's territory beyond the walls of the city.

20. B-P, 1:124, 126–127; 2:205, 207, 210–212, 217, 221; Bowsky, *Finance of the Commune of Siena,* 140–141, 319.

21. The expenditures for both semesters in 1307 came to some 160,085 lire, whereas operating costs for the water system (*bottino* works, cleaning, maintenance, and salaries) totaled 253 lire 11s., or approximately 0.16 percent of the annual budget. In 1341, during the building campaign for the Campo fountain, hydraulic costs rose to about 0.24 percent of the Biccherna's annual expenditures. The Biccherna volumes alone did not represent the commune's entire budget, but the figures do indicate that in spite of the high cost, the public water system consumed a minor fraction of total civic expenditures. Bowsky, *Finance of the Commune of Siena,* app. 1; B-P, 2:175–177, 204–206; Anne Coffin Hanson, *Jacopo della Quercia's Fonte Gaia* (Oxford: Clarendon Press, 1965), 7.

22. Henry Thomas Riley, ed., *Munimenta Gildhallae Londoniensis: Liber Albus, Liber Custumarum, et Liber Horn,* vol. 2, no. 1, RS, no. 12 (London: Longman, Brown, Green, Longmans & Roberts, 1860), 64–66. Riley, *Memorials,* 107, 264–265; *Cal.L-B E,* 220; *Cal.L-B F,* 28–29; *Cal.P&M L. 1323–1364,* 144.

23. Riley, *Memorials,* 264–265.

24. *Cal.L-B H,* 116; *Cal.L-B K,* 318; Charles Lethbridge Kingsford, ed., *Chronicles of London* (Oxford: Clarendon Press, 1905), 188.

25. Harris, *Coventry Leet Book,* 2:517; 3:586.

26. Quotations in Riley, *Memorials,* 521; *Cal.L-B L,* 207. Stat.1251–52, 1.65, 3.200, B-P, 2:292.

27. Bryan Ward-Perkins, *From Classical Antiquity to the Middle Ages: Urban Public Building in Northern and Central Italy,* A.D. 300–850 (Oxford: Oxford Univ. Press, 1984), 140; *Cal.L-B L,* 4.

28. *Cal.Wills,* 1:78; 2:10–11, 15, 32, 37, 200.

29. Lester K. Little, *Religious Poverty and the Profit Economy in Medieval Europe* (London: P. Elek, 1978), 212–213.

30. Attilio Carosi, *Le epigrafi medievali di Viterbo (secc. VI–XV)* (Viterbo: Consorzio per la Gestione delle Biblioteche Comunale degli Ardenti e Provinciale A. Anselmi, 1986), nos. 19, 20, 22, 23, 26, 27, 37.

31. The financing of Dublin's water system reflects the sorts of mixed revenues available for urban water projects. According to the terms of the 1244 writ, the conduit was to be built at the cost of citizens, and damage was to be repaired at the

cost of the king. The income from the rents of certain Dublin properties was applied to conduit maintenance; a proposal made by the "common folk" to the civic officials suggested that fines levied on shops violating closing hours be applied to public works. Henry F. Berry, "The Water Supply of Ancient Dublin," *Journal of the Royal Society of Antiquaries of Ireland* 21 (1890–91): 563–564; *Cal.Dublin,* nos. 24, 34, 87, 106.

32. Benet was mayor in 1409, 1413, and 1418. John Speed, *The History and Antiquity of Southampton,* Southampton Record Society, no. 5 (Southampton: Cox & Sharland, 1909), 26. Colin Platt, *Medieval Southampton: The Port and Trading Community, A.D. 1000–1600* (London: Routledge & Kegan Paul, 1973), 144, 233; George Oliver, ed., *Monasticon Dioecesis Exoniensis* (London: Longman, Brown, Green & Longmans, 1846), 403; *Cal.L-B E,* 228; *Cal.L-B H,* 79, 108; *Cal.L-B K,* 357–358; *Cal.L-B L,* 207; Marjorie Honeybourne, "The Fleet and Its Neighborhood in Early and Medieval Times," *TLMAS* 19 (1947): 58; John Stow, *A Survey of London,* ed. Charles L. Kingsford, 2 vols. (Oxford: Clarendon Press, 1908), 1:16–19, 37, 109–114; 2:41; *Cal.E.M.R.,* 112; *Cal.Wills,* 1:330; 2:175–176, 218–219, 300–301, 307, 324–325, 510, 514.

33. In 1476 John Adam gave four fothers of lead for the conduit and in return for his gift obtained permanent exemption from serving in any office of the corporation. Edward Gillett, and Kenneth A. MacMahon, *A History of Hull* (Oxford: Oxford Univ. Press, 1980), 41–43.

CHAPTER 5. USERS

1. Anne U. White, "Patterns of Domestic Water Use in Low-Income Communities," in *WWHHC,* 108–111; George M. Foster, *Traditional Societies and Technological Change,* 2d ed. (New York: Harper & Row, 1973), 146–174.

2. *Aelfrici Colloquium,* 35, quoted in David Knowles, *The Monastic Order in England: A History of Its development from the Times of St. Dunstan to the Fourth Lateran Council, 943–1216* (Cambridge: Cambridge Univ. Press, 1941), 464 n. *Rule of St. Benedict,* chap. 40; Knowles, *The Monastic Order in England,* 464–465, 717. Judith M. Bennett, *Ale, Beer, and Brewsters in England: Women's Work in a Changing World, 1300–1600* (New York: Oxford Univ. Press, 1996); William Urry, *Canterbury under the Angevin Kings* (London: Athlone Press, 1967), 163.

3. Cum quis de te forte potat, si sit sanus, tunc egrotat, conturbas precordia. Venter tonat, surgit ventus, qui inclusus, non ademptus, multa dat supplicia. Quando venter est inflatus, tunc diversos reddit flattus ex utroque gutture. Salimbene de Adam, *Cronica,* ed. Giuseppe Scalia, 2 vols, Corpus Christianorum, Continuatio Medievalis, no. 125 (Turnhout, Belgium: Brepols, 1998–99), 2:650. English translation: Salimbene de Adam, Joseph L. Baird, Giuseppe Baglivi, and John

Robert Kane, *The Chronicle of Salimbene de Adam,* Medieval and Renaissance Texts and Studies, no. 40 (Binghamton, N.Y.: Center for Medieval and Early Renaissance Studies, 1986), 436.

4. Salimbene de Adam, *Cronica,* 1:364.

5. William Melczer, *The Pilgrim's Guide to Santiago de Compostela* (New York: Italica Press, 1993), 88–90. Lynn Thorndike, *History of Magic and Experimental Science* (New York: Columbia Univ. Press, 1923), 2:133; Rodolfo Lanciani, *The Golden Days of the Renaissance in Rome from the Pontificate of Julius II to That of Paul III* (Boston: Houghton, Mifflin, 1906), 79; R. A. Hall, "The Waterfronts of York," in *Waterfront Archaeology,* ed. G. L. Good, R. H. Jones, and M. W. Ponsford, CBA Research Report no. 74 (London: CBA, 1991), 177–184; Norman Hammond, "Buried Bones Reveal the Risks of Medieval Urban Life," *Times* (London), March 18, 1997.

6. C. T. Flower, ed., *Public Works in Mediaeval Law,* Selden Society, no. 40 (London: Bernard Quaritch, 1923), 2:207–214. Barbara A. Hanawalt, *The Ties That Bound: Peasant Families in Medieval England* (Oxford: Oxford Univ. Press, 1986), 145, 147, 158–159, 271–274. *Cal.Cor.R,* 198–199, 265–266.

7. White, "Patterns of Domestic Water Use," 96–97, 103–105.

8. P. Giuseppe Abate, *S. Rosa da Viterbo, terziaria francescana (1233–1251): Fonti storiche della vita e loro revisione critica* (Rome: Editrice "Miscellanea Francescana," 1952), 131.

9. This is often the only major fracture found in broken pitchers. Manuscript illuminations frequently show the holder of a pitcher supporting the base with one hand while holding the handle with the other; others show jugs carried on the head or the shoulder. Henry Hodges, "The Medieval Potter: Artisan or Artist?" in *Medieval Pottery from Excavations: Studies Presented to Gerald Clough Dunning,* ed. Vera I. Evison, H. Hodges, and J. G. Hurst (London: J. Baker, 1974), 38–39; Michael R. McCarthy and Catherine M. Brooks, *Medieval Pottery in Britain A.D. 900–1600* (Leicester: Leicester Univ. Press, 1988), 111–112, jacket illustration.

10. Catherine of Siena, *The Letters of St. Catherine of Siena,* ed. and trans. Suzanne Noffke, Medieval and Renaissance Texts and Studies, no. 52 (Binghamton, N.Y.: Center for Medieval and Early Renaissance Studies, SUNY Binghamton, 1988), 1:180–181. Cf. John 4:14. B-P, 2:81.

11. *Cal.E.M.R.,* 83.

12. Joseph Thomas Fowler, ed., *Extracts from the Account Rolls of the Abbey of Durham,* 3 vols., Surtees Society, nos. 99, 100, 103 (Durham: Andrews & Co., 1898–1901) 1:9; 2:517, 519, 525–527, 536–537, 543, 546, 550, 554, 559; Henry Charles Coote, "The Ordinances of Some Secular Guilds of London, 1354 to 1496," *TLMAS* 4, no. 1 (1871): 55–58; Felix Barker and Peter Jackson, *The History of London in Maps* (London: Guild, 1990), 19; Alfred Stanley Foord, *Springs, Streams, and Spas of*

London (London: T. F. Unwin, 1910), 273; Lanciani, *The Golden Days of the Renaissance in Rome,* 79–80.

13. Daniel Waley, *Siena and the Sienese in the Thirteenth Century* (Cambridge: Cambridge Univ. Press, 1991), 17–18, 25; Neville Bartlett, ed., *The Lay Poll Tax Returns for the City of York in 1381* (London: A. Brown & Son, 1953), 37, 46, 71; *Cal.Wills,* 1:509, 559.

14. John Stow, *A Survey of London,* ed. Charles L. Kingsford, 2 vols. (Oxford: Clarendon Press, 1908), 1:17. In 1335 the wardens repaired two broken brass taps (*clavorum*). *Cal.L-B F,* 29.

15. Henry Thomas Riley, ed., *Memorials of London and London Life in the XIIIth, XIVth, and XVth Centuries* (London: Longmans, Green & Co., 1868), 225. Roberta J. Magnusson, *Medieval Water Supplies: Hydraulic Technology and Social Organization in England and Italy* (Ph.D diss., Univ. of California, Berkeley; Ann Arbor, Mich.: University Microfilms, 1994), 443–447.

16. London's conduit was leased to private operators in 1367. See chap. 4, p. 120. Stow, *A Survey of London,* 1:17, 190–191; *Cal.L-B H,* 108, 343, 354; Riley, *Memorials,* 264–265, 521; Peter R. V. Marsden, "Archaeological Finds in the City of London, 1963–4," *TLMAS* 21 (1967): 215.

17. Stow, *A Survey of London,* 1:17–19, 37, 109–114, 128, 173, 188–192, 211, 230, 232, 264–268, 284–285, 292–293, 300, 342; 2:3, 11, 18, 21, 34, 38, 40–43, 79, 83; Foord, *Springs, Streams, and Spas of London,* 256 ff.

18. Riley, *Memorials,* 617.

19. Quoted in Foord, *Springs, Streams, and Spas of London,* 274.

20. Lynn bylaws, December 17, 1390, and May 1426. Steven Alsford, ed., "Medieval English Urban History," November 26, 1998 //www.trytel.com/tristan/towns/towns.html; Mary Dormer Harris, ed., *The Coventry Leet Book,* 4 vols., EETS, nos. 134, 135, 138, 146 (London: K. Paul, Trench, Trübner & Co., 1907–13), 1:208, 232; 2:338, 517, 548–549; 3:584, 788, 808–809, 812; William Barclay Parsons, *Engineers and Engineering in the Renaissance* (Cambridge: MIT Press, 1967), 242–243.

21. Quotation in Riley, *Memorials,* 649. White, "Patterns of Domestic Water Use," 107; *Cal.Cor.R.,* 100, 127, 252–253. For those who could afford them, wells were preferred to rivers as sources of water. The bishop of Winchester's palace in Southwark was built along the riverfront: the palace drains discharged into the river, but the water supply came from wells in the courtyard and kitchen. Martha Carlin, "The Reconstruction of Winchester House, Southwark," *LTR* 25 (1985): 33–57.

22. Riley, *Memorials,* 254. *Cal.Cor.R,* 129; *Cal.L-B G,* 206.

23. Coote, "The Ordinances of Some Secular Guilds," 55–58.

24. Riley, *Memorials,* 222–225.

25. Fatal accidents occasionally arose from the practice of watering horses, and

disputes between men watering horses and industrial users of ditches and rivers could become violent. Watering horses seems to have been generally (but not exclusively) a masculine activity. Riley, *Memorials,* 4, 549; Stow, *A Survey of London,* 1:126; Helen M. Cam, ed., *The Eyre of London 14 Edward II A.D. 1321,* vol. 26, nos. 1, 2 of *Year Books of Edward II,* SS, nos. 85, 86 (London: Quaritch, 1968–69), 64, 81; Edwin Brezette DeWindt, ed. and trans., *The Court Rolls of Ramsey, Hepmangrove, and Bury, 1268–1600,* Subsidia Mediaevalia, no. 17 (Toronto: Pontifical Institute of Medieval Studies, 1990), fiche 1, 47; Stat.1251–52, 4.83.

26. Stat.1237–38, caps. 253, 258, 396, 444; Stat.1251–52, 1.55, 57, 60, 61, 3.55, 200, 237; 4.70, 83.

27. Con.1262, 3.82, 88, 197 marginal note, 213; Con.1309, 3.87, 97; B-P, 2:10, 348. The disorder, "caputmorbum vel farcimen," may have been the equine disease known as farcy or glanders.

28. Con.1309, 3.227; B-P, 2:348, 511; *Cal.Cor.R,* 100; Riley, *Memorials,* 648–649; Harris, *Coventry Leet Book,* 2:312, 338.

29. A *lavatoio* basin remains at Siena's Fonte Nuova, but it seems to be a later feature. It may, however, preserve the approximate form of its predecessor; it is fed by water from the main basin. B-P, 2:73, 101, 416; Cecilia Piana Agostinetti, *Fontane a Viterbo: Presenze vive nella città* (Rome: Palombi, 1985), 151.

30. Stat.1237–38, cap. 207; Stat.1251–52, 3.59; Charles de la Roncière, "Tuscan Notables on the Eve of the Renaissance," in *Revelations of the Medieval World,* ed. Georges Duby, trans. Arthur Goldhammer, vol. 2 of *A History of Private Life* (Cambridge: Harvard Univ. Press, Belknap Press, 1988), 209; Waley, *Siena and the Sienese,* 24–25. As modern examples have shown, the social factor in washing clothes can be very important in traditional societies. If public laundries preserve the opportunity for women to maintain social ties with their friends, they will gladly use them, since they provide greater comfort than the riverbank. Attempts to provide women with private washing facilities are less successful: the greater convenience is not considered a sufficient compensation for the ensuing social isolation. Foster, *Traditional Societies,* 112–113.

31. B-P, 2:416; A. Hamilton Thompson, ed., *Visitations of Religious Houses in the Diocese of Lincoln,* 3 vols., Lincoln Record Society, nos. 7, 14, 21 (Horncastle: Lincoln Record Society, 1913–29), 1:74, 98; 2:18, 72, 131–132; 3:308; John Willis Clark, ed., *The Observances in Use at the Augustinian Priory of S. Giles and S. Andrew at Barnwell, Cambridgeshire* (Cambridge: Macmillan & Bowes, 1897), 195; Meredith Parsons Lillich, "Cleanliness with Godliness: A Discussion of Medieval Monastic Plumbing," in *Mélanges à la mémoire du père Anselme Dimier,* ed. Benoît Chauvin, vol. 3, no. 5 (Pupillin, Arbois: B. Chauvin, 1982), 129.

32. Melczer, *The Pilgrim's Guide to Santiago,* 89–90. *HKW,* 1:550; Georges Vigarello, *Concepts of Cleanliness: Changing Attitudes in France since the Middle*

Ages, trans. J. Birrell (Cambridge: Cambridge Univ. Press, 1988), 21. For accidental drownings of bathers in medieval London, see *Cal.Cor.R,* 59, 127, 190; Riley, *Memorials,* 6–7.

33. Quotation in *Cal.P&M L. 1364–1381,* 54. London's stews had a bad reputation and were suppressed in 1417. Baths at hot springs were thought to be medically therapeutic. The commune of Viterbo purchased many parcels of land together with their appurtenances (*piscine* [pools], *alveos* [troughs or channels?], *cursus aquarum* [watercourses], *balneas* [baths], and *aquedotti* [aqueducts]) in a nearby thermal zone in 1293 and 1294. Riley, *Memorials,* 534–535, 647–648; Waley, *Siena and the Sienese,* 158–159; Stat.1236–37, cap. 445; Stat.1251–52, 1.30, 4.146; "Margarita," vol. 2, fols. 101v–106r; vol. 3, fols. 58r–61v, Biblioteca Comunale di Viterbo; B-P, 2:138.

34. The increased availability of water for washing (both personal bathing and the washing of kitchen utensils) reduces the rate of certain diseases. These include the same oral-fecal diseases associated with impure drinking water (which can also be spread by dirty hands or food) and skin and eye infections. David J. Bradley, "Health Aspects of Water Supplies in Tropical Countries," in *WWHHC,* 7, 12–13; Richard G. Feachem, "Water Supplies for Low-Income Communities: Resource Allocation, Planning and Design for a Crisis Situation," in *WWHHC,* 85–86; Charles Homer Haskins, *Studies in the History of Mediaeval Science* (New York: Ungar, 1960), 257; O. G. Tomkeieff, *Life in Norman England* (London: Batsford, 1966), 42; *HKW,* 1:246, 504, 550; 2:697, 926, 934, 974, 998, 1000.

35. Urry, *Canterbury under the Angevin Kings,* 16, 18, 159; Lanfranc, archbishop of Canterbury, *The Monastic Constitutions of Lanfranc,* ed. and trans. David Knowles (London: T. Nelson & Sons, 1951), 10; Clark, *The Observances in Use at the Augustinian Priory of Barnwell,* 195; Joseph Stevenson, ed., *Chronicon Monasterii de Abingdon,* RS, no. 2 (London: Longman, Brown, Green, Longmans & Roberts, 1858), 2:300.

36. *Rule of St. Benedict,* chap. 36; Lanfranc, *The Monastic Constitutions,* 9–10, 26, 62; Lillich, "Cleanliness with Godliness," 129; Gerd Zimmermann, *Ordensleben und Lebensstandard: Die Cura Corporis in den Ordensvorschriften des abendländischen Hochmittelatlers,* Beiträge zur Geschichte des alten Mönchtums und des Benediktineordens, no. 32 (Münster Westfalen: Aschendorff, 1973), 124–126; Jeremiah F. O'Sullivan, ed., *The Register of Eudes of Rouen,* trans. Sydney M. Brown (New York: Columbia Univ. Press, 1964); F. M. Powicke, ed. and trans., *The Life of Ailred of Rievaulx by Walter Daniel,* Medieval Classics (London: Thomas Nelson & Sons, 1950), 25, 34; Louis J. Lekai, *The Cistercians: Ideals and Reality* (Kent, Ohio: Kent State Univ. Press, 1977), 374–375.

37. Yorkshire case quoted in A. R. Martin, *Franciscan Architecture in England,* British Society of Franciscan Studies, no. 18 (Manchester: Manchester Univ.

Press, 1937), 120. Saint Denis rules quoted in Clark, *The Observances in Use at the Augustinian Priory of Barnwell,* lxxxv. Salimbene de Adam, *Cronica,* 2:837; Zimmermann, *Ordensleben und Lebensstandard,* 433; Robert Willis, "The Architectural History of the Conventual Buildings of the Monastery of Christ Church in Canterbury," *Arch.Cant.* 7 (1868): 184.

38. Peter Fergusson, "The Twelfth-Century Refectories at Rievaulx and Byland Abbeys," in *Cistercian Art and Architecture in the British Isles,* ed. Christopher Norton and David Park (Cambridge: Cambridge Univ. Press, 1986), 178–179.

39. Clark, *The Observances in Use at the Augustinian Priory of Barnwell,* 155, 195; Willis, "The Architectural History of Christ Church in Canterbury," 62, 188–189; Lanfranc, *The Monastic Constitutions,* 92–93; Zimmermann, *Ordensleben und Lebensstandard,* 126–128; William Thorne, *William Thorne's Chronicle of St. Augustine's Abbey Canterbury,* trans. A. H. Davis (Oxford: Basil Blackwell, 1934); Thompson, *Visitations of Religious Houses,* 1:2; 2:22.

40. Thérèse and Mendel Metzger, *Jewish Life in the Middle Ages: Illuminated Hebrew Manuscripts of the Thirteenth to the Sixteenth Centuries* (New York: Alpine Fine Arts Collection, 1982), 101; Frederick J. Furnivall, ed., *The Babees Book,* EETS, o.s., no. 32 (London: N. Trübner & Co., 1868), 5, 8, 13, 283, 303, 321–323; R. W. Chambers, ed., *A Fifteenth-Century Courtesy Book,* EETS, no. 148 (Oxford: K. Paul, Trench, Trübner & Co., 1914), 11–14; Georges Duby, ed., *Revelations of the Medieval World,* trans. Arthur Goldhammer, vol. 2 of *A History of Private Life* (Cambridge: Harvard Univ. Press, Belknap Press, 1988), 232, 364–366, 376, 502, 524–525, 600–609.

41. Stat.1237–38, cap. 207; Con.1262, 3.179; Con.1309, 3.107–109; 5.146; B-P, 2:17, 26, 48; Harris, *Coventry Leet Book,* 1:208; "Lo Statuto dell'Arte dei Macellaj del 1384," fol. 18, marginal note, Biblioteca Comunale di Viterbo.

42. Bartlett, *Lay Poll Tax Returns for the City of York,* 7–9; Urry, *Canterbury under the Angevin Kings,* 121–124.

43. The Brook Street channels were lined with wood or stone. Martin Biddle, ed., *Winchester in the Early Middle Ages: An Edition and Discussion of the Winton Domesday,* Winchester Studies, no. 1 (Oxford: Clarendon Press, 1976), 284, 345, 434, 438, tables 47, 48; "Margarita," vol. 2, fols. 101v–106; vol. 3, fol. 58; Stat.1236–37, caps. 247, 382; Stat.1251–52, 1.30, 34; 4.193.

44. The term *guazzatoio* would seem to indicate that it was a watering trough for animals, but in this case its function appears to overlap that of the *abbeveratoio.* In a few instances the two words seem to be used interchangeably for the same basin, but in other cases they clearly refer to two distinct basins—Fonte Branda and Fonte Nobili had both. It was forbidden to wash wool, put leather, or throw lime and hair (presumably wastes generated by the preparation of hides) in Fonte Branda's *guazzatoio.* Luciano Banchi and Filippo-Luigi Polidori, eds., *Statuti senesi scritti in*

volgare ne'secoli XIII e XIV, 3 vols. (Bologna: G. Romagnoli, 1863–77), 1:172; 2:349; B-P, 1:50; 2:26, 27, 49, 132, 453; Con.1262, 3.176, 179 and marginal note; Con.1309, 3.105, 107–108; 4.10; Sandra Tortoli, "Per la storia della produzione laniera a Siena nel Trecento e nei primi anni del Quattrocento," *Bullettino Senese di Storia Patria* 82–83 (1977): 222.

45. Henry Thomas Riley, trans., *Liber Albus: The White Book of the City of London* (London: Richard Griffin, 1861), 280; Helena M. Chew and William Kellaway, eds., *London Assize of Nuisance 1301–1431: A Calendar,* London Record Society, no. 10 (London: London Record Society, 1974), xxv.

46. *Cal.Cor.R,* 167–168; *Cal.L-B B,* 277; Clemens Kosch, "Wasserbaueinrichtungen in hochmittelalterlichen Konventanlagen Mitteleuropas," in *WIM,* 96.

47. Joseph Thomas Fowler, ed., *Rites of Durham,* Surtees Society, no. 107 (Durham: Andrews & Co., 1903), 85; Salimbene quotation in Salimbene de Adam et al., *The Chronicle of Salimbene de Adam,* 577–578. Lanfranc, *The Monastic Constitutions,* 79, 117; see Salimbene de Adam, *Cronica,* 2:857–858, for the Latin version of Salimbene's words.

48. Ernest L. Sabine, "Latrines and Cesspools of Mediaeval London," *Speculum* 9, no. 3 (1934): 306–309. *Cal.E.M.R,* 247; Derek Portman, *Exeter Houses: 1400–1700* (Exeter: Exeter Univ., 1966), 15; P. E. Jones, "Whittington's Longhouse," *LTR* 23 (1972): 27–34.

49. Colin Platt and Richard Coleman-Smith, *Excavations in Medieval Southampton 1953–1969,* 2 vols. (Leicester: Leicester Univ. Press, 1975), 1:34, 293, 302–304; George Lambrick, "Further Excavations on the Second Site of the Dominican Priory, Oxford," *Oxoniensia* 50 (1985): 195–198, 207; Jean E. Mellor and T. Pearce, *The Austin Friars, Leicester,* CBA Research Report no. 35 (London: CBA, 1981), 15, 24–25, 41–42; P. J. Drury, "Chelmsford Dominican Priory: The Excavation of the Reredorter, 1973," *Essex Archaeology and History* 6 (1974): 50; Stewart Cruden, "Glenluce Abbey: Finds Recovered during Excavations," *Transactions, Dumfriesshire and Galloway Natural History and Antiquarian Society* 29 (1950–51): figs. 7–9, 181–182; McCarthy and Brooks, *Medieval Pottery in Britain,* 115–116. A domestic drain in Bristol had become blocked when several vessels thrown into the drain became jammed against a misaligned stone. K. J. Barton, "Excavations at Back Hall, Bristol, 1958," *TBGAS* 79, no. 2 (1960): 262.

50. Since the Canterbury drain was flushed only with rainwater and with the wastewater from the piped system, it suffered from low flow levels and would have required more frequent cleaning than drains flushed by diverted watercourses. Harold Brakspear, "Bardney Abbey," *Arch.J.* 79 (1922): 53; L. F. Salzman, *Building in England down to 1540* (Oxford: Clarendon Press, 1952), 279–80; Urry, *Canterbury under the Angevin Kings,* 57, 173. Sabine, "Latrines and Cesspools," 310.

51. Salzman, *Building in England,* 284–285; Salimbene de Adam, *Cronica,* 1:388–392, 449.

52. Quotations in Colin Platt, *Medieval Southampton: The Port and Trading Community, A.D. 1000–1600* (London: Routledge & Kegan Paul, 1973), 181; and Salimbene de Adam et al., *The Chronicle of Salimbene de Adam,* 56. See Salimbene de Adam, *Cronica,* 1:114, for the Latin text of the Detsalve account. James Greig, "The Investigation of a Medieval Barrel-Latrine from Worcester," *JAS* 8, no. 3 (1981): 281.

53. B-P, 2:122; Ignazio Ciampi, ed., *Cronache e statuti della città di Viterbo* (Bologna: A. Forni, 1976), 35, 396–397; Cesare Pinzi, *Storia della città di Viterbo,* 4 vols., Biblioteca Istorica della Antica e Nuova Italia, no. 195 (Rome: Tip. della Camera dei Deputati, 1887–99), 3:345–349.

54. Stat.1237–38, cap. 253; Stat.1251–52, 1.57.

55. Modern studies of immigrants arriving in cities in Mexico and Ghana have found that most stay with relatives or fellow villagers when they first arrive. Lester K. Little, *Religious Poverty and the Profit Economy in Medieval Europe* (London: P. Elek, 1978), 23–25, 28–29; Foster, *Traditional Societies,* 49.

56. Quotation in *Cal.P&M L. 1364–1381,* 2, 15. Christopher Dyer, *Standards of Living in the Later Middle Ages: Social Change in England c. 1200–1520* (Cambridge: Cambridge Univ. Press, 1989), 191; P. V. Addyman, "The Archaeology of Public Health at York, England," *World Archaeology* 21, no. 2 (1989): 244–263; Platt and Coleman-Smith, *Excavations in Medieval Southampton,* 1:34; *Cal.E.M.R,* 40.

EPILOGUE

1. James Gairdner, ed., *Letters and Papers, Foreign and Domestic of the Reign of Henry VIII,* vol. 13, pt. 1, 2d ed. (Vaduz, Lichtenstein: Kraus Reprint, 1965), no. 1342. Terrence James, "Excavations at Carmarthen Greyfriars, 1983–1990," *MA* 41 (1997): 154.

2. "Cronica S. Petri Erfordensis Moderna," in *Supplementa Tomorum XVI–XXV,* MGH Scriptores, no. 30, pt. 1 (Stuttgart: Anton Hiersemann, 1964), 442; Walter Minchinton, *Life to the City: An Illustrated History of Exeter's Water Supply from the Romans to the Present Day* (Newton Abbot, Eng.: Devon Books, 1987), 18; Jim Gould, "The Twelfth-Century Water-Supply to Lichfield Close," *Ant.J.* 56 (1976): 77; *VCH Worcester,* 4:396; Walter George Bell, *The Great Fire of London in 1666* (London: John Lane, 1920), 59–60, 93, 127.

3. Monique Wabont, *Maubuisson au fil de l'eau. Les réseaux hydrauliques de l'abbaye du XIIIe au XVIIIe siècle,* Notice d'archéologie du Val-d'Oise, no. 3 (Saint-Ouen-l'Aumone: Service Départmental d'Archéologie du Val-d'Oise, 1992), 33–34, 37–39.

4. *L&P H VIII,* vol. 13, pt. 2, no. 674. Frederick C. Jones, *Bristol's Water Supply and Its Story* (Bristol: St. Stephen's Bristol Press, 1946), 13; Robert Willis, *The Architectural History of the University of Cambridge and of the Colleges of Cam-*

bridge and Eton (Cambridge: Cambridge Univ. Press, 1886), 2:430, 627–632; William Reader, *The History and Antiquities of the City of Coventry* (Coventry, 1810), 50–51; *St. Mary's Conduit, Lincoln,* Lincolnshire Museums Information Sheet, Archaeology Series, no. 19 (Lincoln: Lincolnshire County Council, Lincolnshire Museums, 1980).

5. The city withstood royal pressure for some time but eventually caved in. They permitted the bishop his pipe, but the tap was restricted to the size of a grain of vetch, and they reserved the right to cut the supply if the public fountain ran short. Pericle Perali, "L'acquedotto medievale orvietano. Studio storico e topografico," in *La città costruita: Lavori pubblici e immagine in Orvieto medievale,* by Lucio Riccetti (Florence: Le Lettere, 1992), 251, 273–275, figs. 9, 11; William Barclay Parsons, *Engineers and Engineering in the Renaissance* (Cambridge: MIT Press, 1967), 241–247.

6. In the model for the growth of large technological systems formulated by Thomas Hughes, the components of expanding systems that lag behind and impede the full achievement of the system's goal ("reverse salients") may come to be perceived as "critical problems" requiring a solution. Solutions of critical problems often result in systems growth; but if this proves insufficient, a new system may emerge and exist (at least for a while) in competition with the older system. Thomas P. Hughes, "The Evolution of Large Technological Systems," in *The Social Construction of Technological Systems: New Directions in the Sociology and History of Technology,* ed. Wiebe E. Bijker, Thomas P. Hughes, and Trevor J. Pinch (Cambridge: MIT Press, 1987), 71–76.

7. Modern practice recommends a head of 25 feet for wood pipes. This will ensure that the wood reaches saturation pressure quickly and remains completely saturated, which retards decay. Wood was a cheaper material than lead, and wood pipes were much longer than earthenware pipes. Since there were fewer joints, installation and maintenance were less labor-intensive. National Tank and Pipe Company, *A Handbook of Wood Pipe Practice* (Portland, Oreg.: National Tank & Pipe Co., 1938), 7–8; Bernard Rudden, *The New River: A Legal History* (Oxford: Clarendon Press, 1985).

8. The diffusion of the new technology cluster and the switch to private sponsorship seem to be more characteristic of northern Europe than Italy and are almost certainly linked to larger social and economic transformations. I am hesitant about making broader claims, however, as the subject awaits its own full-scale study. Thorkild Schiøler, *Roman and Islamic Water-Lifting Wheels* (Odense, Denmark: Odense Univ. Press, 1973); John Peter Oleson, *Greek and Roman Mechanical Water-Lifting Devices: The History of a Technology* (Toronto: Toronto Univ. Press, 1984); Ahmad Y. al-Hassan and Donald R. Hill, *Islamic Technology: An Illustrated History* (Cambridge: Cambridge Univ. Press, 1986), 37–52; Niklaus Schnitter, *Die*

Geschichte des Wasserbaus in der Schweiz (Oberbözberg, Switz.: Olynthus, 1992), 58–60; Albrecht Hoffmann, "Frühe Trinkwasser-Pumpwerke in Hessen," *Wasser und Boden* 40, no. 4 (1988): 203–206; Jan Werth, "Ursachen und technische Voraussetzungen für die Entwicklung der Wasserbehälter," in *Die Architektur der Förder- und Wassertürme,* ed. Bernhard and Hilla Becher (Munich: Prestel Verlag, 1971), 330–335.

9. Kenneth J. Knoespel, "Gazing on Technology: *Theatrum Mechanorum* and the Assimilation of Renaissance Machinery," in *Literature and Technology,* ed. Mark L. Greenberg and Lance Schachterle, Research in Technology Studies, no. 5 (Bethlehem, Pa.: Lehigh Univ. Press, 1992), 99–124; Mariano di Jacopo detto il Taccola, *Liber Tertius de Ingeneis ac Edifitiis non Usitatis,* ed. J. H. Beck (Milano: Polifilo, 1969); Frank Prager and Gustina Scaglia, *Mariano Taccola and His Book De Ingeneis* (Cambridge: MIT Press, 1972).

10. Royal Commission on Water Supply, *Report of the Commissioners Presented to Both Houses of Parliament by Command of Her Majesty,* Reports from Commissioners: 1868–69, no. 22 (London: HMSO, 1869), pt. 3, sec. 96; F. Williamson, "George Sorocold, of Derby: A Pioneer of Water Supply," *Journal of the Derbyshire Archaeological and Natural History Society* 57 (1936): 65–93.

11. Bell, *Great Fire of London,* 249; T. F. Reddaway, *The Rebuilding of London after the Great Fire* (London: Edward Arnold, 1951), 37, 56, 79, 168, 282–283; Williamson, "George Sorocold," 83; J. A. Hassan, "The Growth and Impact of the British Water Industry in the Nineteenth Century," *Economic History Review* 38 (1985): 531–547.

12. W. M. Frazer, *A History of English Public Health, 1834–1939* (London: Baillière, Tindall & Cox, 1950), 63–67, 135–140. *Report by the General Board of Health on the Supply of Water to the Metropolis Presented to Both Houses of Parliament by Command of Her Majesty,* Reports from the Commissioners, 1850, no. 22 (London: HMSO, 1850), 287, 319–322. Among the more compelling arguments for a new supply is that "the saving in tea from the use of soft water may be estimated at about one-third of the tea consumed by the metropolis." Voting with their feet during the typhoid epidemic of 1905, Lincoln's citizens lined up with buckets to draw water from the old, spring-fed Greyfriars' conduit. *St. Mary's Conduit, Lincoln,* 5.

13. Royal Commission on Water Supply, *Report* (1869), pt. 3, sec. 95; pt. 5.4, sec. 246.

Notes on Sources

The primary evidence for medieval water systems comes in three main forms: textual, representational, and physical. Information can be gleaned from many different types of documents as well as from inscriptions, plans, art, archaeology, and standing remains. The main hurdle I faced in my own research was tracking down references that were widely scattered. The problem has been reduced in recent years by several excellent regional studies, which summarize the textual and archaeological evidence and provide essential regional bibliographies. Here I give a brief overview of the types of primary sources I found to be most informative and present a selection of secondary works. For more exhaustive documentation, I advise readers to refer to my Ph.D. dissertation, *Medieval Water Supplies: Hydraulic Technology and Social Organization in England and Italy* (Univ. of California, Berkeley, 1994; Ann Arbor, Mich.: University Microfilms, 1994).

Medieval charters and licenses are perhaps the most abundant class of documents relating to conduits, and they are the most important sources for the analysis of land and resource acquisition. The terms of the agreements may include incidental technical information (such as the type of pipe or the dimensions of the conduit house), clauses pertaining to maintenance, and compensation for damages. If place names are identifiable, they may be used to determine the general route of the conduit. Charters can also provide specific dates or at least allow the date to be bracketed within the terms of office of known individuals. The researcher must remember, however, that the actual construction may have taken place considerably later (if at all). In England royal licenses were issued to many (but by no means all) conduit builders and were recorded in the Patent Rolls. These licenses provide information similar to that found in charters, and the same caveat about dates applies. Charters can be found in monastic cartularies and civic archives, and many collections have been published, although often they are left untranslated. For conduit licenses, see the *Calendar of the Patent Rolls.* (For the volumes up to 1317–21, "conduit" appears in the index; for later years it does not.)

Narrative sources occasionally contain valuable information. A detailed description of the construction of Waltham Abbey's water system appears to be the work of an interested observer (probably one of the canons) who witnessed the work personally. It is accompanied by a second account and a plan of the head of the conduit. The Latin texts of the Waltham narratives are available in R. A. Skelton and P. D. A. Harvey, *Local Maps and Plans from Medieval England* (Oxford: Clarendon Press, 1986), and Rosalind Ransford's *The Early Charters of the Augustinian Canons of Waltham Abbey, Essex 1062–1230* (Woodbridge, Eng.: Boydell Press, 1989). The latter work also includes the charters pertaining to the conduit. A detailed topographical description of the London Franciscan conduit is contained in Charles Lethbridge Kingsford's *The Grey Friars of London: Their History with the Register of Their Convent and an Appendix of Documents* (Aberdeen, Scotland: Aberdeen Univ. Press, 1915). Chronicles and annals sometimes note the construction of water systems and may include dates, names of sponsors, financial backers, the master in charge of the works, and incidental technical information. Biographical narratives, such as the *vitae* (lives) of saints or the *gestae* (deeds) of notable figures, may also include incidental references to water supplies. Works commemorating the deeds of abbots and other monastic officials will often note their sponsorship of water systems, as may obituaries and funerary inscriptions. The construction of a water system was considered so important an achievement that it might be given pride of place in such memorials, even for men like Canterbury's Prior Wibert or Bury Saint Edmund's Abbot Samson, who are known to have also sponsored other major building projects.

Civic records, which sometimes include petitions and proposals for hydraulic projects, are a particularly useful source for adoption decisions and system administration. They may include information about the anticipated (and actual) costs of projects, the means used to raise money, and the officials involved in various aspects of waterworks administration and construction. Not all medieval civic records are published, though many are preserved in local archives. A comprehensive collection of extracts from municipal records relating to Siena's water system was compiled by Fabio Bargagli-Petrucci and comprises volume 2 of his *Le fonti di Siena e I loro aquedotti* (Siena: L. S. Olschki, 1903). This is probably the most important published collection of documents for a civic water system, even if the transcriptions are not always accurate. Giusta Nicco Fasola's *La Fontana di Perugia* (Rome: Libreria dello Stato, 1951) and Pericle Perali's "L'acquedotto medievale orvietano," in *La città costruita: Lavori pubblici e immagine in Orvieto medievale,* by Lucio Riccetti (Florence: Le Lettere, 1992), also contain collections of relevant excerpts from civic records. A wealth of miscellaneous information about London's conduit and rivers can be found in the *Calendar of Letter-Books Preserved among the Archives of the Corporation of the City of London at the Guildhall,* ed. Reginald R. Sharpe (London: Francis, 1899–1912), and in Henry Thomas Riley's collection of sources in *Memorials of London and London Life in the XIIIth, XIVth, and XVth Centuries* (London: Longmans, Green & Co., 1868). For other civic systems, see *The Coventry Leet Book,* ed. Mary Dormer Harris (London: K. Paul, Trench, Trübner & Co., 1907–13), and the publications of local record societies, such as the Southampton Record Society or the Bristol Record Society.

Financial records can be rich mines of information for the construction, maintenance, administration, and costs of hydraulic projects. They often include the names of master craftsmen and (more rarely) workmen, along with occupational specialties and wages. It is sometimes possible to calculate the amount of time and the numbers of workers needed for specific operations. In addition, financial accounts may contain information concerning the purchase and transport of construction materials and tools. They may also record the names, wages, and duties of officials in charge of the day-to-day operation of a water system. I particularly recommend *The Accounts of the Fabric of Exeter Cathedral, 1279–1353,* ed. and trans. Audrey M. Erskine (Torquay, Eng.: Devonshire Press, 1981–83), *Extracts from the Account Rolls of the Abbey of Durham,* ed. Joseph Thomas Fowler (Durham: Andrews & Co., 1898–1901), and the city of Siena's Biccherna registers. Some volumes of the latter have been published as the *Libri dell'entrata e dell'uscita del comune di Siena: Detti della Biccherna* (Pubblicazioni degli Archivi di Stato; Rome: Ministero dell'interno, 1961), and most of the relevant entries are included in *Le fonti di Siena.*

Law codes and legal commentaries include provisions governing water rights, hydraulic structures, and sanitation and can shed light on underlying attitudes toward hygiene. Ordinances governing many aspects of civic water supplies appear in civic statutes. These may include orders for hydraulic structures to be built, modified, cleaned, and repaired. They also contain information about administration and about regulations for users. Individual statutes may be repeated in subsequent redactions of a code, so dating their initial appearance can be a problem; furthermore, it is not always easy to judge how effective they were. The Siena statutes are particularly informative on all aspects of the water system. See Lodovico Zdekauer, ed., *Il constituto del comune di Siena dell'anno 1262* (Milan: U. Hoepli, 1897); *Il costituto del comune di Siena volgarizzato nel* MCCCIX–MCCCX (Siena: L. Lazzeri, 1903); and Luciano Banchi and Filippo-Luigi Polidori, eds., *Statuti senesi scritti in volgare ne'secoli XIII e XIV e pubblicati secondo I testi del R. Archivio di stato in Siena* (Bologna: G. Romagnoli, 1863–77).

Monastic customaries (*consuetudines*) sometimes include provisions concerning the use, administration, and maintenance of water systems. As in the case of urban statutes, it is not always easy to determine how closely reality matched their prescriptions. Some texts incorporate entire sections copied from the customaries of other houses, so references to a water system must be treated with caution. Provisions from customaries relating to monastic hygiene can be found in Gerd Zimmermann, *Ordensleben und Lebensstandard: Die Cura Corporis in den Ordensvorschriften des abendländischen Hochmittelatlers* (Münster Westfalen: Aschendorff, 1973).

Court records, such as eyre rolls or local court rolls, record the disputes, infractions, and abuses associated with water use. These are good sources of information about users and abusive waste-disposal practices. A litany of complaints and abuses can be found in the *London Assize of Nuisance 1301–1431: A Calendar,* ed. Helena M. Chew and William Kellaway (London: London Record Society, 1974). Accidental deaths were more typically associated with rivers and wells than with fountains, but coroners' records do contain much incidental information about users, and they show that traditional water sources were still employed alongside complex water systems. Wills occasionally include benefactions for conduits or other public works and are indicative of emerging attitudes toward urban infrastructures.

Cartographic evidence for water systems is rare but of critical importance. Four medieval waterworks plans are known, all from England. The first two are an elaborate plan and a simpler diagram of the water system at Christ Church, Canterbury. Both of these drawings are thought to be nearly contemporary with the initial construction of the water system (1153–67). A thirteenth-century diagram shows the head of Waltham Abbey's conduit. The fourth is a fifteenth-century plan of the London Charterhouse water system (which shows, incidentally, parts of two other conduits). The plans include details of individual hydraulic components, providing invaluable evidence for the configurations of complete systems. Skelton and Harvey's *Local Maps and Plans from Medieval England* contains excellent color reproductions of all four plans, along with transcriptions of the accompanying texts.

Postmedieval records and plans can shed light on the subsequent history of medieval water systems. In Britain, inventories of the goods of religious houses were drawn up at the Dissolution, and some include hydraulic components such as lead pipes and bronze taps. The Christ's Hospital financial accounts and their waterworks plan of 1676 help illuminate the London Greyfriars' system, which the hospital obtained following the suppression of the friary. Some postmedieval narrative sources, such as John Stow's *A Survey of London,* ed. Charles L. Kingsford (Oxford: Clarendon, 1908), and *The Itinerary of John Leland,* ed. Lucy Toulmin Smith (Carbondale: Southern Illinois Univ. Press, 1964), contain information about medieval water systems.

Works of art and literature may include representations of hydraulic structures, such as fountains, or activities, such as bathing. Some of these images may be closely modeled on real structures and everyday life, but others are fanciful or symbolic devices that are more likely to reflect artistic conventions and fashions than actual technology and behavior. I have hesitated to draw inferences from these imaginative depictions, since I am not an expert in the disciplines of art history and literary criticism. Nonetheless, the evidence is there for those who know how to handle it.

Surviving physical evidence for medieval water systems can supplement and help clarify the imprecise descriptions and technical terminology found in documentary sources. A number of fountains and conduit houses still exist, and some are still functioning. Such survivors are excellent sources of technical information; in addition, the symbolism of their decorations may provide clues to the attitudes and political agendas of their builders. When more precise evidence is lacking, the style can provide a rough guide to the date. Less tangibly, a visit to a monastic laver still standing in its cloister fountain house or to a fountain still gracing a civic piazza makes it easier to understand hydraulic structures in their architectural and spatial contexts.

The proliferation of medieval excavations, along with improvements in field techniques and scientific analysis, has greatly increased the quantity and quality of archaeological evidence for the structural components of water systems. Archaeologists are also undertaking detailed analyses of environmental evidence, which can shed light on waste disposal and sanitation practices. The analysis of hydraulic components can give clues to manufacturing and construction techniques and the overall hydraulic efficiency of systems, although published reports are not consistent in providing technical details. Most countries have a national journal devoted to medieval archaeology, such as Britain's *Medieval Archaeology* or Italy's *Archeologia Medievale,* which presents articles, book reviews, and short summaries of current projects. Many excavation reports appear only in local journals, whereas some medieval excavations are published in journals with broader chronological parameters, such as *The Antiquaries Journal*. Some early excavators, such as W. H. St. John Hope, Harold Brakspear, and Philip Norman, seem to have been particularly interested in medieval waterworks, and their reports are still worth consulting.

Monique Wabont's *Maubuisson au fil de l'eau. Les réseaux hydrauliques de l'abbaye du XIIIe au XVIIIe siècle* (Saint-Ouen-l'Aumone: Service Départmental d'Archéologie du Val-d'Oise, 1992) is a model archaeological publication of a complete water system. Most excavations, however, reveal only segments of medieval systems; there are different survival rates for organic versus inorganic materials and above-ground versus subterranean structures. Components made of metal are less likely to survive in situ than components made of materials such as clay or stone, which were not worth looting or recycling. Archaeological dating can be imprecise, and it is risky to use the physical characteristics of hydraulic components as dating evidence, since synchronic and diachronic variations are not fully understood. Some systems can be dated by their stratigraphic relationships to other features and buildings, and in some cases documentary evidence for a water system can help establish a date for archaeological remains.

The secondary literature on medieval water supplies is not vast, but it is growing. An essential work is the Frontinus-Gesellschaft's *Die Wasserversorgung im Mittelalter* (Mainz am Rhein: P. Von Zabern, 1991), which contains a collection of excellent regional studies and is lavishly illustrated. *Working with Water in Medieval Europe,* edited by Paolo Squatriti (Leiden: Brill, forthcoming), contains a collection of regional surveys covering multiple aspects of hydraulic technology, including urban and ecclesiastical water systems, canals, irrigation, mills, fishing, flood control, and land drainage. Occasionally, entire volumes in a series, such as *Mélanges de l'école française de Rome Moyen Age,* vol. 104, no. 2 (1992) and *L'eau au moyen âge,* Senefiance, no. 15 (Aix-en-Provence, Marseille: Publications du CUER MA, Univ. de Provence, 1985), are devoted to the topic of water.

For early medieval Italy, see Paolo Squatriti, *Water and Society in Early Medieval Italy, A.D. 400–1000* (Cambridge: Cambridge Univ. Press, 1998). Bryan Ward-Perkins's *From Classical Antiquity to the Middle Ages: Urban Public Building in Northern and Central Italy, A.D. 300–850* (Oxford: Oxford Univ. Press, 1984) contains excellent chapters on water systems in late antiquity and the early Middle Ages.

Monastic water supplies are perhaps the best studied medieval hydraulic systems. The conference papers presented in Léon Pressouyre and Paul Benoit, eds., *L'hydraulique monastique: Milieux, réseaux, usages,* Rencontres à Royaumont, no. 8 (Grâne, France: Créaphis, 1996), give an excellent overview of current research across Europe. I also recommend Meredith Parsons Lillich, "Cleanliness with Godliness: A Discussion of Medieval Monastic Plumbing," *Mélanges à la mémoire du père Anselme Dimier,* ed. Benoît Chauvin, vol. 3, no. 5 (Pupillin, Arbois:

B. Chauvin, 1982). For England, see the works of C. James Bond, particularly "Water Management in the Rural Monastery," in *The Archaeology of Rural Monasteries,* ed. R. Gilchrist and H. Mytum (Oxford: British Archaeological Reports, 1989); and "Water Management in the Urban Monastery," in *Advances in Monastic Archaeology,* ed. Gilchrist and Mytum (Oxford: British Archaeological Reports, 1993). There are excellent chapters on water management in Glyn Coppack's *English Heritage Book of Abbeys and Priories* (London: Batsford, 1990) and J. Patrick Greene's *Medieval Monasteries* (Leicester: Leicester Univ. Press, 1992).

For urban water supplies, see W. C. Wijntjes, "The Water Supply of the Medieval Town," *Rotterdam Papers* 4 (1982); Dietrich Lohrmann, "Neues über Wasserversorgung und Wassertechnik im Mittelalter," *Deutsches Archiv* 48, no.1 (1992); Jürgen Sydow, ed., *Städtische Versorgung und Entsorgung im Wandel der Geschichte,* Stadt in der Geschichte, no. 8 (Sigmaringen, Germany: Jan Thorbecke, 1981); and Bernd Herrmann, ed., *Mensch und Umwelt im Mittelalter* (Stuttgart: Deutsche Verlags-Anstalt, 1986). André Guillerme's *The Age of Water: The Urban Environment in the North of France,* A.D. *300–1800* (College Station: Texas A&M Univ. Press, 1988) has little to say about piped water systems but is very good on intramural canal networks. See also his paper "Puits, aqueducs et fontaines: L'alimentation en eau dans les villes du nord de la France, Xe–XIIIe siècles," in *L'eau au moyen âge* (1985).

Local histories of individual towns may mention medieval water systems and can provide a good introduction to local primary sources. For some cities, more in-depth hydraulic studies are available. For Italian towns, see Nevio Basezzi and Bruno Signorelli, *Gli antichi acquedotti di Bergamo* (Bergamo: Comune di Bergamo, 1992); Gianni Botturi and Remo Pareccini, *Antichi acquedotti del territorio bresciano* (Milan: Edizioni Et, 1991); Armando Schiavo, *Acquedotti romane e medioevali* (Naples: F. Giannini, 1935) (for Salerno and Vietri sul Mare); Cecilia Piana Agostinetti, *Fontane a Viterbo: Presenze vive nella città* (Rome: Palombi, 1985); Daniela Monacchi, "L'acquedotto Formina di Narni," *Bollettino d'Arte* 39–40 (1986); and *Siena e l'acqua: Storia e immagini di una città e delle sue fonti,* ed. Vinicio Serino (Siena: Nuova Immagine Editrice, 1997). On Siena see also the many works of Duccio Balestracci. Alfred Stanley Foord's *Springs, Streams, and Spas of London* (London: T. F. Unwin, 1910) includes a good deal of information on the medieval conduits. Laure Beaumont-Maillet's *L'eau à Paris* (Paris: Hazan, 1991) has a chapter on the city's medieval water supply.

On the medieval urban environment and sanitation, see Derek J. Keene, "The Medieval Urban Environment in Documentary Records," *Archives* 16, no. 70 (1983), and the articles on London's butchers, latrines, and city cleaning by Ernest L. Sabine, which appear in *Speculum,* vols. 8, 9, and 12 (1933, 1934, 1937). Ronald E. Zupko and Robert A. Laures, *Straws in the Wind: Medieval Urban Environmen-*

tal Law: The Case of Northern Italy (Boulder: Westview Press, 1996), contains a detailed study of sanitation statutes and is a good introduction to Italian civic law codes. For archaeological approaches, see *Environmental Archaeology in the Urban Context,* ed. R. Hall and H. K. Kenward (London: Council for British Archaeology, 1982).

Some of the more general works on the history of hydraulic technology and engineering that include sections on the Middle Ages are F. W. Robins's *The Story of Water Supply* (London: Oxford Univ. Press, 1946); R. J. Forbes, "Hydraulic Engineering and Sanitation," in *A History of Technology,* vol. 2, ed. Charles Singer, E. J. Holmyard, and A. R. Hall (Oxford: Clarendon Press, 1956); Norman A. F. Smith's *Man and Water: A History of Hydro-Technology* (New York: Scribner, 1975); and Donald R. Hill, *A History of Engineering in Classical and Medieval Times* (London: Croom Helm, 1984). William Barclay Parsons, *Engineers and Engineering in the Renaissance* (Cambridge: MIT Press, 1967), includes information on late medieval waterworks. L. F. Salzman's *Building in England down to 1540* (Oxford: Clarendon Press, 1952) contains a good deal of information about the medieval construction trades.

Medieval technology was heavily influenced by the technological traditions of the ancient world. The literature on ancient water systems is vast. For recent studies with up-to-date bibliographies, see the Frontinus-Gesellschaft's series Geschichte der Wasserversorgung, which includes *Wasserversorgung im Antiken Rom* (Munich: R. Oldenbourg, 1982) and *Die Wasserversorgung Antiker Städte* (Mainz am Rhein: P. von Zabern, 1987–88). A. Trevor Hodge's *Roman Aqueducts & Water Supply* (London: Duckworth, 1992) is an excellent study of technical aspects of Roman systems. On hydraulic technology in medieval Spain and elsewhere in the medieval Islamic world, see the works of Thomas F. Glick, especially *Irrigation and Hydraulic Technology: Medieval Spain and its Legacy* (Aldershot, Eng.: Variorum, 1996), and Ahmad Yusuf Hasan and Donald R. Hill, *Islamic Technology: An Illustrated History* (Cambridge: Cambridge Univ. Press, 1986).

Recent theoretical approaches to the study of technological change have been summarized by John M. Staudenmaier in *Technology's Storytellers: Reweaving the Human Fabric* (Cambridge: MIT Press, 1985). The collection of papers in *The Social Construction of Technological Systems,* ed. Wiebe E. Bijker, Thomas P. Hughes, and Trevor J. Pinch (Cambridge: MIT Press, 1987), includes Thomas Hughes's model for "The Evolution of Large Technological Systems" as well as articles by proponents for the social construction of technology (SCOT). Floyd Shoemaker and Everett M. Rogers have constructed highly detailed models for the diffusion of innovations by the statistical analysis of a multitude of case studies. The most recent synthesis is Rogers's *Diffusion of Innovations,* 4th ed. (New York: Free Press, 1995). For an anthropological perspective on technological change, see

George M. Foster, *Traditional Societies and Technological Change,* 2d ed. (New York: Harper & Row, 1973). Some recent attempts to introduce modern water systems in traditional societies are explored in Richard G. Feachem, Michael McGarry, and Duncan Mara, eds., *Water, Wastes, and Health in Hot Climates* (London: Wiley, 1977).

For the study of individual medieval water systems, the best place to begin is in the appropriate local history collection and archive. Many cities and counties also have their own archaeological societies and museums, where it may be possible to consult original field records from excavations, examine artifacts firsthand, and find out about unpublished sites. For comparative studies like this one, a copyright library is essential. I was lucky enough to spend several years in Oxford, where I trawled through the justifiably famous collections of the Bodleian Library and also took advantage of a vast collection of archaeological publications in the Ashmolean Museum's reading room. Increasingly, information about medieval waterworks is becoming available over the Internet, though the Web sites generally lack source citations and should be used with caution.

Index

Page numbers in italics refer to illustrations.